Secondary Cohomology Operations

John R. Harper

Graduate Studies in Mathematics
Volume 49

American Mathematical Society
Providence, Rhode Island

2000 *Mathematics Subject Classification*. Primary 55-01, 55S05, 55S10, 55S20, 55S45, 55P99.

Abstract. This textbook develops the theory of secondary cohomology operations for singular cohomology theory and makes applications in the form of explicit calculations. The treatment is intended for graduate students with a knowledge of basic algebraic topology including exposure to the Steenrod operations. The subject is developed in terms of elementary constructions from general homotopy theory. Among the applications, there are proofs of the Hopf invariant one theorems for all primes. The final chapter treats the theory of Massey-Peterson fibrations in order to enlarge the scope of the basic theory through universal examples. This chapter also includes twisted operations and Cartan formulas.

Library of Congress Cataloging-in-Publication Data

Harper, John R., 1941–
Secondary cohomology operations / John R. Harper.
p. cm. — (Graduate studies in mathematics, ISSN 1065-7339 ; v. 49)
Includes bibliographical references and index.
ISBN 0-8218-3198-4 (acid-free paper) ISBN 0-8218-3270-0 (softcover)
1. Homology theory. I. Title. II. Series.
QA612.3 .H36 2002
514′.23–dc21 2002023236

∞ The paper used in this book is acid-free and falls within the guidelines
established to ensure permanence and durability.
Visit the AMS home page at URL: http://www.ams.org/

10 9 8 7 6 5 4 3 2 1 07 06 05 04 03 02

In memory of Alexander Zabrodsky (1936–1986).

Contents

Preface

Secondary cohomology operations are one of the tools available which bear on questions left unresolved by primary operations. This book develops this specialized topic in terms of elementary concepts from general homotopy theory. The special circumstances of their applications mean that secondary operations are often found embedded in detailed computations and other technicalities. It has been known for some time that the subject can be set up in elementary terms. Perhaps more recent is the understanding that there are systematic strategies for making calculations which can also be developed in the same elementary framework. For the author, that understanding emerged in joint work with Alex Zabrodsky.

The first six chapters of this book develop the subject along the lines alluded to above. This work takes us through a proof of the Hopf invariant one theorem of J. F. Adams for the prime 2 and Liulevicius, Shimada and Yamanoshita for odd primes. Our proofs are in the spirit of those works but do not rely on calculations of the cohomology of universal examples. Moreover, by applying the elegant method employed by Shimada and Yamanoshita on the Steenrod algebra, the relation leading to the factorization of the appropriate Steenrod operation by secondary operations is worked out without working through the cohomology of the Steenrod algebra.

Besides the Hopf invariant one theorem, many other results about classical secondary operations are presented in the first six chapters. Notable among these is Browder's evaluation of higher order Bocksteins on p-th powers.

Our approach to the subject is through the idea of secondary compositions. This is an old idea, having great success in the hands of Barratt and Toda, among others, in the study of the homotopy groups of spheres. Many people have realized that secondary compositions supply a description of secondary cohomology operations and Spanier published an account of the basic theory. Nevertheless, it usually transpires that to make calculations, one must rely on ad hoc information in many cases or on advanced methods developed for the Adams spectral sequence.

In our work on finite H-spaces, Zabrodsky and I were confronted with the evaluation of a certain p-th order operation which was inaccessible by any method we knew at that time. We were able to resolve our problem by using the Milnor filtration where a space is regarded as the classifying space of its loop space. We realized that this method gave an alternative path through most of the literature where secondary operations were calculated and Alex gave a series of lectures in this vein for a workshop held in Barcelona. The idea that a textbook devoted to a similar treatment might be useful comes from the fact that the method still finds uses and the belief that the subject has both elegance and scope.

The table of contents indicates what may be found. Here I want to make some marginal comments on the material. I expect that readers of this book are familiar with the Steenrod algebra and its uses. For many people this means knowing the basic properties through the Adem relations and Milnor's structure theorem for the Hopf algebra. The first chapter interweaves a geometric discussion of primary operations with a summary discussion of features of the Steenrod operations. Also present is one of the systematic strategies for calculations. It is an argument first given by Adem to study compositions. I call it the Adem argument to indicate its general nature.

On the algebraic side, chapter one contains a new proof of a theorem due to A. Negishi concerning certain left multiplications in the Steenrod algebra. This result is used at the prime 2 to give the same argument that Shimada and Yamanoshita give at odd primes for the relation factoring certain Steenrod operations through secondary operations. Naturally the level of detail here exceeds that of a summary discussion. I have also included a largely unnecessary discussion of the ideal in the Steenrod algebra consisting of operations annihilating classes of a fixed dimension. This material is included because it simplifies some of my early work on the subject.

The overall purpose of the first chapter is to have the Steenrod algebra and the cohomology of Eilenberg-Mac Lane spaces ready for use in subsequent work. I have not tried to develop these topics, even in sketch form.

Our treatment of secondary operations deviates from the approach found in most of the literature. I will to try to delineate the differences. Typically, a secondary operation is produced in the cohomology of a universal example. This approach, where an element is picked out of a module, does not come with means for evaluation in specific cases. Our approach is to represent the cohomology class by a map known as a colifting defined in homotopy theoretic terms from the same data defining the operation. Then an evaluation of the operation is represented by a secondary composition. This simple geometric description entails a general formula for making calculations. In the literature, it is known as the Peterson-Stein formula or compatibility with connecting homomorphisms in Cartan's treatment. In elementary terms, the method is simply an adjoint relationship appearing in diagrams which can be recognized in many calculations. In the work with Zabrodsky mentioned above, we found exactly these patterns presented by the Milnor filtration. For us, the key feature provided by this filtration is an analysis of essential maps which become null-homotopic upon looping. Our direct hold on this phenomenon is another place where our treatment differs from most of the literature. There, the phenomenon is encoded in terms of splittings of universal examples as spaces, but not as H-spaces.

Chapter two is a bridge to the point of view dominating our development. Chapter three presents the basic geometric theory including our version of the Peterson-Stein formula and our use of the Milnor filtration. I have tried to separate those elements which appear to be general from those particular to secondary operations. Chapter four develops the basic theory of secondary operations. Except for the language, there is no difference between the results of our development and the traditional ones found in Adams' paper, the Cartan seminar, or the book by Mosher and Tangora. In chapter four we also indicate how our theory applies to operations of order higher than secondary operations. However, we do not develop this aspect in a systematic way. It is my opinion that technical matters get in the way of understanding unless one already has a good hold on the secondary situation. Moreover, I am unaware of Milnor filtration type information for the universal examples that serve higher order operations except for the case of higher order Bocksteins.

Chapter five presents several examples where the Milnor filtration comes into the story. All these calculations appear as direct applications of the basic strategies, strategies arising from recognition of a common pattern described in chapter two.

In particular, we have the evaluation of Adams and Liulevicius-Shimada-Yamanoshita operations in the cohomology of complex projective space. Moreover, the algebraic part of the decomposition formula is produced in this chapter. Thus the means to settle Hopf invariant one are present but the denouement is delayed until the next chapter.

Readers wishing to follow a connected account of the Hopf invariant one theorems can do so by leaving chapter three after Prop. 3.5.3 ((a) if

$p = 2$ is preferred) omitting the material in chapter four after subsection 4.2.8 (exercises 4.2.3–5 may be omitted) and going to section 5.3 (5.4 for odd primes).

Chapter six contains the Hopf invariant one results. The first part includes classical background material following lectures of John Moore. The second part assembles previous work to finish the proofs for the cases left open by the classical work.

As may be inferred, we do not base our work on the cohomology of universal examples, but it would be perverse to ignore this topic. Chapter seven is devoted in part to the work of Massey and Peterson which provides the most comprehensive hold on the cohomology structure of the spaces arising in the classical theory. The geometric work of earlier chapters takes its place in this theory, in particular, in the discussion of the Hopf algebra structure and the principal action.

Chapter seven also includes a discussion of twisted operations and Cartan formulas for secondary operations. I know that unrestrained glee is inappropriate to a sober preface, but let me say that I was pleased to find that the material of chapters three and four could sustain the discussion of these topics.

It is my belief that the ideas developed in this text can continue to be of use in homotopy theory. The book includes many examples and exercises with the intention that the reader will work through these as the principal means to understand the subject. Some of these examples, especially in chapter seven, are important for applications, notably in obstruction theory. However, I have only supplied references and have not tried to sketch the applications themselves. In fact, an excellent exercise is to look at the references and rework the relevant parts. Who knows, maybe that will lead to further understanding of these classical topics.

I mentioned earlier that the point of view for this book grew out of work with Alex Zabrodsky. I like to think that Alex would have enjoyed sharing authorship of this book. I would like to think that his influence here is undiminished, but I know that is not true. This book in dedicated in memoriam to Alex as thanks for the pleasure of working with him and for his profound contributions to homotopy theory.

Acknowledgement

It is a pleasure to thank people who have helped me with this book. Foremost is Joan Robinson, who prepared this book from a handwritten manuscript. Both John Moore and Haynes Miller supplied me with detailed comments on the early sections. Joe Roitberg attended a graduate course I offered in 1990 and Fred Cohen and Moore did the same in 1994. Their presence allowed me to try out various ways to present the material and they were not reluctant to tell me when something did not work. The graduate students in those courses also deserve my thanks for their attendance. Sam Gitler has listened patiently to my explanations and Joe Neisendorfer arranged for a semester leave of absence during which I wrote the first draft. Because it will please them, let me acknowledge two grandlings, Alana (1995) and Erika (1998). Their antics provided welcome diversion from the regimen imposed on my writing.

Review of Primary Operations

1.1. Primary cohomology operations

The topological work in this book takes place in the category $\mathcal{K}_*$ of compactly generated spaces with non-degenerate base point, as discussed in [**117**]. We begin with the definition of a primary cohomology operation.

Definition 1.1.1. Let $H^n(\quad;\pi)$ and $H^q(\quad;G)$ be the singular cohomology functors from the category of topological pairs and continuous maps to the category of sets and functions, with n and q positive. A *primary cohomology operation* θ *of type* (π, n, G, q) *is a natural transformation from* $H^n(\quad;\pi)$ *to* $H^q(\quad;G)$.

Thus for any pair (X, Y) we have a function

$$\theta(X,Y) : H^n(X,Y;\pi) \to H^q(X,Y;G)$$

and for any map of pairs $f : (X, Y) \to (W, Z)$, we have

$$\theta(X,Y) \circ f^* = f^* \circ \theta(W,Z)$$

where f^* is the map induced on cohomology. We take $n, q > 0$ in our definition to exclude examples of type $(0, 0, G, 0)$ where $G \neq 0$.

1.1.1. Consequences of naturality. $\theta(X,Y)$ is a pointed map of pointed sets. Since the cohomology functors take values in the category of abelian groups, there is a distinguished element 0 in each group $H^n(X, Y; \tau)$. Then it follows from naturality and the values for the cohomology of a point that θ is a pointed map of pointed sets.

$\theta(X, Y)$ **is the zero map for** $q < n$**.** Starting with a CW pair, we observe that the q-th cohomology of the pair maps monomorphically into the q-th cohomology of the q-skeleton of the pair. By naturality, this composition factors through 0. For arbitrary pairs, the result follows by use of the CW approximation theorem.

Semi-additivity.

A cohomology operation need not be additive. This point is emphasized by regarding the cohomology of a pair only as pointed sets and not as abelian groups in (1.1). However, there is a limited form of additivity, which we call "semi-additivity", and is defined as follows. Let (X, Y) be an NDR pair and $\delta : \bar{H}^{n-1}(Y; \pi) \to H^n(X, Y; \pi)$ be the connecting homomorphism from reduced cohomology. For this discussion, we regard the source and target of δ as abelian groups. Then for classes u in $H^n(X, Y; \pi)$ and v in $\bar{H}^{n-1}(Y; \pi)$ we have the *semi-additivity formula*

$$\theta(u + \delta v) = \theta(u) + \theta(\delta v) \ ,$$

where θ is an abbreviation for $\theta(X, Y)$ with source $H^n(X, Y; \pi)$.

The proof is elementary after we set up some notation. Let I be the unit interval. Form S^1 by identifying 0 with 1 and take the image of 0 to be the base point. Let ΣX be the reduced suspension $S^1 \wedge X$. Let $T_0Y, T_1Y, \Sigma Y$, and T_0X be respectively the subspaces of ΣX determined by $[0, \frac{1}{2}] \times Y$, $[\frac{1}{2}, 1] \times Y$, $S^1 \wedge Y$, and $[0, \frac{1}{2}] \times X$. We identify X with the image of $\{\frac{1}{2}\} \times X$ in ΣX and likewise for Y.

We have the following diagram of pairs and inclusions, where $*$ is the basepoint:

$$\begin{array}{lccccccl}
\text{(a)} & (X, Y) & \subset & (X \cup T_1Y, T_1Y) & \supset & (X \cup T_1Y,\ *) & & \\
 & \cap & & \cap & & \cap & & q \\
\text{(b)} & (T_0X, Y) & \subset & (T_0X \cup T_1Y, T_1Y) & \supset & (T_0X \cup T_1Y,\ *) & & \\
 & \cup & & \cup & & \cup & & i \\
\text{(c)} & (T_0Y, Y) & \subset & (\Sigma Y, T_1Y) & \supset & (\Sigma Y,\ *)\ . & &
\end{array}$$

The vertical maps on the right are named for use in the argument. Each horizontal inclusion and each inclusion from row (c) to row (b) induces an isomorphism in cohomology which commutes with θ. Let $\bar{u}$ in $H^n(X \cup T_1Y; \pi)$ be the image of u under the isomorphisms from row (a). Let $\bar{v}$ in $H^n(T_0X \cup T_1Y; \pi)$ be the image of $\delta_b(v)$, the connecting homomorphism for row (b), under the isomorphisms from row (b). Then $u + \delta_a(v)$ is mapped to $\bar{u} + q^*\delta_b(v)$ by isomorphisms that commute with θ. Thus it is enough to prove

$$\theta(\bar{u} + q^*\bar{v}) = \theta(\bar{u}) + \theta(q^*\bar{v}) \ .$$

To achieve this, we introduce a map

$$c : X \cup T_1 Y \to (X \cup T_1 Y) \vee \Sigma Y$$

obtained by pinching the image of $\{3/4\} \times Y$ to a point in $X \cup T_1 Y$. The map c is the identity on X and is given by the formula

$$(t, y) \to \begin{cases} (2t - \frac{1}{2}, y) \text{ in } X \cup T_1 Y & \text{for } \frac{1}{2} \leq t \leq 3/4 \,, \\ (4t - 3, y) \text{ in } \Sigma Y & \text{for } 3/4 \leq t \leq 1 \,. \end{cases}$$

We have the following array of inclusion and projection maps,

$$X \cup T_1 Y \underset{p_1}{\overset{j_1}{\rightleftarrows}} X \cup T_1 Y \vee \Sigma Y \underset{j_2}{\overset{p_2}{\rightleftarrows}} \Sigma Y$$

with $p_1(\Sigma Y) = *$ in $T_1 Y$ and $p_2(X \cup T_1 Y) = *$ in ΣY.

These maps induce a direct sum decomposition for the cohomology of $X \cup T_1 Y \vee \Sigma Y$. Furthermore, we have the equation

$$q \simeq i \circ p_2 \circ c \,.$$

So we can write $\bar{u} + q^* \bar{v} = c^*(p_1^* \bar{u} + p_2^* i^* \bar{v})$. By naturality, application of j_1^* and j_2^* respectively to $\theta(p_1^* \bar{u} + p_2^* i^* \bar{v})$ yields $\theta(\bar{u})$, $\theta(i^* \bar{v})$. Hence $\theta(p_1^* \bar{u} + p_2^* i^* \bar{v}) = p_1^* \theta(\bar{u}) + p_2^* \theta(i^* \bar{v})$ by the direct sum decomposition. Combining this fact with the naturality of θ, we have

$$\begin{aligned} \theta(\bar{u} + q^* \bar{v}) &= \theta(c^*(p_1^* \bar{u} + p_2^* i^* \bar{v})) \\ &= \theta(\bar{u}) + \theta(q^* \bar{v}) \,. \end{aligned}$$

Remark. The semi-additivity property first appears in [**93**] and is also discussed in exposé 9 of [**21**].

Remark. An argument similar to the one given here shows that *$\theta(X)$ is additive if X is a co-H-space.* The details are left as an exercise.

1.1.2. Stable cohomology operations. Let $\{\theta_n | n \geq 1\}$ be a sequence of cohomology operations of type $(\pi, n, G, n+i)$ for a fixed positive integer i. We call such a sequence a *stable cohomology operation of degree i* provided

$$\delta \theta_n = \theta_{n+1} \delta$$

for each $n \geq 1$. More precisely, we require that for each pair (X, Y), the following diagram commutes:

$$\begin{array}{ccc} H^n(Y; \pi) & \xrightarrow{\delta} & H^{n+1}(X, Y; G) \\ \Big\downarrow{\scriptstyle \theta_n(Y)} & & \Big\downarrow{\scriptstyle \theta_{n+1}(X,Y)} \\ H^{n+i}(Y; \pi) & \xrightarrow{\delta} & H^{n+i+1}(X, Y; G) \end{array} \qquad \text{for each } n \geq 1 \,.$$

We call the individual θ_n the *components* of a stable operation. It follows from the semi-additivity property applied to the pair consisting of Y and a cone on Y, that the *components of a stable operation are additive.*

Remark. The *Mayer-Vietoris suspension* isomorphism, for non-empty spaces,

$$\Delta^* : \bar{H}^{n-1}(Y;\pi) \to H^n(\Sigma Y;\pi)$$

is defined using the spaces and maps from the semi-additivity argument row (c). Then a stable operation is one whose components commute with suspension

$$\theta_{n+1} \circ \Delta^* = \Delta^* \circ \theta_n \ .$$

Remark. Our Δ^* has the opposite sign of that used in [**117**]. The reason for our choice appears in Proposition 1.3.1.

1.2. Steenrod operations and the Steenrod algebra

1.2.1. Steenrod operations. There is a stable operation Sq^i having components of type $(Z/2, n, Z/2, n+i)$. We suppress the indexing of components and write

$$Sq^i : H^n(X,Y;Z/2) \to H^{n+i}(X,Y;Z/2) \ .$$

These operations enjoy the following properties.

(1) Sq^0 is the identity.

(2) Zero property. If $n < i$, then Sq^i is the zero map.

(3) Squaring property. If $n = i$, then Sq^n is the cup product

$$Sq^n x = x^2 \ .$$

(4) Cartan formula. On cup products, Sq^i satisfies the equation

$$Sq^i(xy) = \sum_{i=j+k} (Sq^j x)(Sq^k y) \ .$$

(5) Sq^1 is the Bockstein connecting homomorphism induced by

$$0 \to Z/2 \to Z/4 \to Z/2 \to 0 \, .$$

This is not a complete list of properties we wish to display, but we can take a break from recitation to make some calculations.

The mod 2 cohomology algebras of projective spaces can be determined from their homology by using Poincaré Duality. The results follow:

real projective space	$H^*(RP^n;Z/2) = Z/2[x]/(x^{n+1})$	with degree $x = 1$,
complex projective space	$H^*(CP^n;Z/2) = Z/2[y]/(y^{n+1})$	with degree $y = 2$,
quaternionic projective space	$H^*(HP^n;Z/2) = Z/2[z]/(z^{n+1})$	with degree $z = 4$,
Cayley projective plane	$H^*(W;Z/2) = Z/2[w]/(w^3)$	with degree $w = 8$.

With the properties enumerated above, the following calculations can be made:

$$Sq^i x^k = \binom{k}{i} x^{k+i} \,,$$

$$Sq^{2i} y^k = \binom{k}{i} y^{k+i} \,,$$

$$Sq^{4i} z^k = \binom{k}{i} z^{k+i} \,.$$

In these formulas, the binomial coefficients are reduced mod 2 and the conventions

$$\binom{n}{i} = 0 \text{ if } i < 0 \text{ or } i > n \,,$$

$$\binom{n}{0} = 1 \text{ if } n \geq 0$$

are observed. The reader can prove these equations by induction on $k + i$ and using the Pascal identity

$$\binom{k-1}{i} + \binom{k-1}{i-1} = \binom{k}{i} \,.$$

The stability property for Sq^2 combined with its value in the cohomology algebra of CP^2 yields that Sq^2 is non-zero in the cohomology module for any suspension of CP^2. Consequently, these suspensions do not have the homotopy type of a bouquet of spheres. Historically, this example was among the first where the Steenrod operations were decisive in distinguishing among homotopy types. The following exercises concern situations where knowledge of the Steenrod operations is not enough to make a distinction. They will serve as test cases for later material.

Exercise 1.2.1. In the cohomology of CP^3 show that y^3 is not in the image of any Steenrod operation of positive degree. Thus ΣCP^3 and $\Sigma CP^2 \vee S^7$ are not distinguished by Steenrod operations.

Exercise 1.2.2. Let X be the quotient CP^6/CP^3. Establish a short exact sequence (coefficients are $Z/2$ and $q > 0$)

$$0 \to H^{2q}(X) \to H^{2q}(CP^6) \to H^{2q}(CP^3) \to 0 \,.$$

Prove that for any $n > 0$, Sq^n is the zero map on $H^8(X)$ and Sq^2 maps $H^{10}(X)$ is isomorphically to $H^{12}(X)$. Thus X and $S^8 \vee \Sigma^8 CP^2$ are not distinguished by Steenrod operations.

Exercise 1.2.3. Let S be the algebra $Z/2[w]/(w^4)$ with degree $w = 8$. Show that w^3 is not in the image of any Steenrod operation of positive degree. Thus Steenrod operations do not rule out the possibility of a space T with mod 2 cohomology algebra S.

Next we discuss a more elaborate calculation. Write L_n for the n-fold Cartesian product of RP^∞ with itself. Then the mod 2 cohomology algebra of L_n is the polynomial algebra on one-dimensional generators,

$$H^*(L_n; Z/2) = Z/2[x_1, x_2, \dots, x_n]$$

where x_i is the image of the generator for the i-th factor of L_n under the map induced on cohomology by projection. Since the operations are stable, expressions like $Sq^a Sq^b$ make sense as compositions. We introduce certain compositions and then describe their behavior in the cohomology of L_n.

Definition 1.2.4. Let $I = (i_1, i_2, \dots, i_k, 0, \dots)$ be a sequence of non-negative integers with only finitely many non-zero entries. We write $Sq(I)$ for the composition

$$Sq^{i_1} Sq^{i_2} \cdots Sq^{i_k} .$$

We say I is *admissible* provided $i_k \geq 1$ and

$$i_j \geq 2 i_{j+1}$$

for each entry of I. By fiat, $I = (0, \dots)$ is admissible as well. We call the sum $i_1 + \cdots + i_k$ the *degree* of I.

Next we describe some of the behavior of $Sq(I)$ in the cohomology of L_n for admissible I. Let $\sigma_j = \sigma_j(x_1, \dots, x_n)$ be the j-th elementary symmetric polynomial in the $x_1, \dots, x_n$. In particular, we have the product

$$\sigma_n = x_1 \cdots x_n .$$

The natural ordering of integers $j_1 > j_2$ induces an ordering on the σ_j,

$$\sigma_{j_1} > \sigma_{j_2} \Leftrightarrow j_1 > j_2$$

and hence there is a lexicographic order on monomials in the σ_j. Thus, for monomials expressed as products in decreasing order,

$$\sigma_{j_1} \cdots \sigma_{j_a} > \sigma_{k_1} \cdots \sigma_{k_b}$$

provided $j_1 > k_1$, or if $j_1 = k_1$, then $j_c > k_c$ where c is the first place where they differ.

Proposition 1.2.5. Let I be admissible of degree $\leq n$. Then

$$Sq(I)\sigma_n = \sigma_n \cdot Q_I$$

where $Q_I = \sigma_{i_1}\sigma_{i_2} \dots \sigma_{i_k} +$ terms of lower order with $I = (i_1, i_2, \dots, i_k, 0 \cdots)$.

Proof. We first establish that $Sq^i\sigma_n = \sigma_n\sigma_i$ for $i \leq n$ by induction on n. We have

$$\begin{aligned} Sq^i\sigma_n = &x_1 Sq^i\sigma_{n-1}(x_2,\dots,x_n) + x_1^2 Sq^{i-1}\sigma_{n-1}(x_2,\dots,x_n) \\ =&x_1\sigma_{n-1}(x_2,\dots,x_n)\sigma_i(x_2,\dots,x_n) + x_1^2\sigma_{n-1}(x_2,\dots,x_n)\sigma_{i-1}(x_2,\dots,x_n) \\ =&\sigma_n(\sigma_i(x_2,\dots,x_n) + x_1\sigma_{i-1}(x_2,\dots,x_n)) \\ =&\sigma_n\sigma_i\ . \end{aligned}$$

Next we argue by induction on the length of I. First observe, as a consequence of degrees, that each summand of $Sq^i\sigma_j$, with $i < j$, is dominated by σ_{i+j} ,

$$Sq^i\sigma_j \leq \sigma_{i+j} \text{ if } i < j$$

and

$$Sq^j\sigma_j = \sigma_j^2 < \sigma_{2j}\ .$$

Write $I = (i_1, J)$. Then we have

$$\begin{aligned} Sq(I)\sigma_n &= Sq^{i_1}\sigma_n Q_J \\ &= Sq^{i_1}(\sigma_n\sigma_{i_2}\cdots\sigma_{i_k} + \text{ lower terms}) \\ &= \sigma_n\sigma_{i_1}\cdots\sigma_{i_k} + \sum_{m=0}^{i_1-1}\sigma_n\sigma_m Sq^{i_1-m}(\sigma_{i_2}\cdots\sigma_{i_k}) + \text{ lower terms.} \end{aligned}$$

The summand with $i_1 - m = i_2 - 1$ supplies the highest order term. But $i_1 > 2i_2 - 1$ by admissibility. □

1.2.2. Steenrod algebra. For compositions $Sq^a Sq^b$ with $0 < a < 2b$ we have:

(1) The *Adem relations*

$$Sq^q Sq^b = \Sigma\binom{b-1-t}{a-2t} Sq^{a+b-t} Sq^t\ .$$

The sum may be regarded as over all integers t, with the conventions on binomial coefficients stated in (2.1). Thus non-zero summands appear for values of t satisfying

$$0 \leq t \leq a/2\ .$$

Informally, the *Steenrod algebra* is the graded algebra over $Z/2$ generated by Sq^i, $i \geq 1$, in grade i and subject to the Adem relations. Formally, take the tensor algebra $\underline{\mathcal{A}}$ generated by Sq^i and divide out by the two sided ideal in $\underline{\mathcal{A}}$ generated by the elements

$$R(a,b) = Sq^a \otimes Sq^b - \Sigma\binom{b-1-t}{a-2t} Sq^{a+b-t} \otimes Sq^t\ .$$

We write $\mathcal{A}$ for this quotient algebra of $\underline{\mathcal{A}}$. If $I = (i_1, i_2, \ldots, i_k)$ is a sequence of positive integers, we write

$$Sq(I) \text{ or } Sq^{i_1} Sq^{i_2} \cdots Sq^{i_k}$$

for the composition which is the image of

$$Sq^{i_1} \otimes \cdots \otimes Sq^{i_k} \text{ from } \underline{\mathcal{A}} .$$

If the entries of I involve some arithmetic, we shall use the notation

$$Sq(i_1)Sq(i_2) \cdots Sq(i_k)$$

for $Sq(I)$.

Calculation of the coefficients in the Adem relation is often made using the *Lucas formula.* Let p be a prime and write the p-nary expansions of integers a, b as

$$a = \sum_{i \geq 0} a_i p^i , \quad b = \sum_{i \geq 0} b_i p^i$$

where the coefficients are between 0 and $p - 1$. Then

$$\binom{a}{b} \equiv \prod_i \binom{b_i}{a_i} \quad \bmod p .$$

This formula is helpful to establish:

(2) *As an algebra $\mathcal{A}$ is generated by Sq^{2^k}, $k \geq 0$. Moreover, Sq^{2^k} does not decompose as a product in $\mathcal{A}$.*

Proof. Observe that for $m \geq 1$,

$$Sq(2^k)Sq(2^{k+1}m) = Sq(2^{k+1}m + 2^k) + \text{ decomposable terms.}$$

Hence the first statement follows by induction on degree.

If there is a non-trivial decomposition

$$Sq(2^k) = \Sigma a_j Sq^j , \; j < 2^k ,$$

then we can evaluate both sides on $H^{2^k}(RP^\infty; Z/2)$. But

$$Sq^j x^{2^k} = 0$$

while the left side acts non-trivially. □

(3) *Cartan basis for $\mathcal{A}$. In grade n, $\mathcal{A}$ has a basis consisting of all "admissible monomials" $Sq(I)$ where I is admissible of degree n.*

We sketch the proof. The first step is to prove that admissible monomials span. For this, define the *moment* of I, written $m(I)$ as

$$m(I) = \sum_{j \geq 1} j i_j , \quad I \text{ not necessarily admissible.}$$

Then one observes that application of an Adem relation to a non-admissible monomial expresses it as a sum of monomials, each with smaller moment. Thus an induction is possible. Linear independence follows from Prop. 1.2.5.

Remark. The Cartan basis may be established without the use of the Adem relations. This is done by Serre [**101**]. Then the method of Prop. 1.2.5 may be used to establish the relations.

1.2.3. Milnor's Theorem on the structure of the Steenrod algebra. A global description of the Steenrod algebra was achieved by Milnor in [**81**]. The concept of a Hopf algebra plays a key role in this development. We sketch the steps and refer to [**81**] or [**109**] for details.

(1) *The map of generators*

$$\psi(Sq^i) = \sum_{i=j+k} Sq^j \otimes Sq^k$$

extends to a homomorphism of algebras

$$\psi : \mathcal{A} \to \mathcal{A} \otimes \mathcal{A} .$$

The Steenrod algebra has an augmentation and at most one generator in each degree. Equipped with ψ, $\mathcal{A}$ becomes a Hopf algebra. Since $\mathcal{A}$ is of finite type, a global description of $\mathcal{A}$ as a Hopf algebra can be achieved by such a description for the dual Hopf algebra $\mathcal{A}^*$.

Define elements ξ_i in $\mathcal{A}^*$ of degree $2^i - 1$ by the requirements

$$\langle \xi_i \, , \; Sq(2^{i-1}, 2^{i-2}, \dots, 1) \rangle = 1$$

and

$$\langle \xi_i \, , \; Sq(I) \rangle = 0 \quad \text{otherwise for admissible monomials.}$$

(2) *As a Hopf algebra, $\mathcal{A}^*$ is isomorphic to the polynomial algebra*

$$Z/2[\xi_1, \xi_2, \dots]$$

with degree $\xi_i = 2^i - 1$. The coproduct

$$\varphi^* : \mathcal{A}^* \to \mathcal{A}^* \otimes \mathcal{A}^*$$

is given by

$$\varphi^*(\xi_i) = \sum_{i=j+k} \xi_j^{2^k} \otimes \xi_k + \xi_i \otimes 1 + 1 \otimes \xi_i \, .$$

For later use, we recall some of the constructions made in the proof of Milnor's theorem. Let $I = (i_1, i_2, \dots, i_s, 0, \dots)$ be a sequence of non-negative integers, in which only a finite number of entries are non-zero. We say I has *length* s provided $i_s \neq 0$ and $i_n = 0$ for all $n > s$. We write $\ell n(I) = s$. We write the *degree* of I as deg (I) and define it by

$$\deg(I) = \sum_{i \geq 1} i_j \ .$$

We say that I is *admissible* if Defn. 1.2.4 holds. For admissible monomials, we can define the *excess*, written $ex(I)$, by

$$ex(I) = \sum_{j \geq 1} (i_j - 2i_{j+1})$$

When it is defined, excess satisfies the equation

$$\deg(I) + ex(I) = 2i_1 \ .$$

We set up a similar bookkeeping notation for monomials in the ξ_i. For clarity, we now let $J = (j_1, \dots, j_t, 0 \dots)$ denote a typical sequence of the kind considered above. The term length has been defined. Two new terms associated with J are its *multiplicity*, written $m(J)$, and defined by

$$m(J) = \sum_{r \geq 1} j_r \ ,$$

and $d(J)$ given by

$$d(J) = \sum_{r \geq 1} (2^r - 1) j_r \ .$$

We write $\xi(J)$ for the monomial

$$\xi(J) = \xi_1^{j_1} \cdots \xi_t^{j_t} \ ,$$

provided that the length of J is t. Then the degree of $\xi(J)$ is given by $d(J)$.

We now compare the Cartan basis for $\mathcal{A}$ with the basis of monomials in the polynomial algebra on the ξ_i. Let I be an admissible monomial of length s. Construct a monomial J of length s by setting

$$j_r = i_r - 2i_{r+1}$$

where i_r is the r-th entry in I. In addition to having the same length, I and J satisfy the equations

$$\deg(I) = d(J) \text{ and } ex(I) = m(J) \ .$$

When J is produced from I in this manner, we write $J = I'$. On the other hand, starting with a sequence J of length t, we construct I of length t by setting

$$\begin{aligned} i_1 &= j_1 + 2j_2 + \cdots + 2^{t-1}j_t \\ i_2 &= j_2 + 2j_3 + \cdots + 2^{t-2}j_t \\ &\vdots \\ i_t &= j_t \end{aligned}$$

and set the remaining entries of I equal to 0. It is clear that I is admissible and the equations above are satisfied for the pair I, J. When I is produced from J in this manner, we write $I = J''$.

The two priming operations form an inverse pair of functions and establish that the dimensions of $\mathcal{A}^*$ and the polynomial algebra on the ξ_i are equal in each degree.

The evaluation pairing satisfies the following equation. Let I be admissible of length s and $J = I'$, then

$$\langle \xi(J),\ Sq(I) \rangle = 1 \ .$$

Next, we impose a certain partial ordering on the monomials $\xi(J)$. This ordering is called the "lexicographic ordering from the right." Let J_1, J_2 be two sequences with the same degree. We say that

$$J_1 < J_2 \ ,$$

provided:

(a): $\ell n(J_1) < \ell n(J_2)$;

or if their lengths are equal, say t ,

(b): t-th entry of J_1 is smaller than the t-th entry of J_2;

or if those agree,

(c): the sequences $J_{1,t}$ and $J_{2,t}$ obtained respectively from J_1 and J_2 by changing the t-th entry in each to 0 satisfy

$$J_{1,t} < J_{2,t} \ .$$

The double prime operation is used to transfer this ordering to an ordering of admissible sequences with common degree. Then for $I_1' = J_1 < J_2$,

$$\langle \xi(J_2),\ Sq(I_1) \rangle = 0 \ .$$

Informally, the incidence pairing is represented by a triangular matrix. This concludes our recall of constructions from Milnor's proof.

Next we make some comments about the incidence relations. First of all, the Cartan basis is not dual to the basis of monomials. This is seen as soon as it can happen, in degree 3,

$$\begin{array}{rcc} Sq^2Sq^1 & 1 & 1 \\ Sq^3 & 1 & 0 \\ & \xi_1^3 & \xi_2 \ . \end{array}$$

Nevertheless, Sq^n in the Cartan basis is dual to ξ_1^n in the monomial basis, as these are the terms of lowest order.

Our second remark concerns the use of excess to order the Cartan basis. Let I_1, I_2 be admissible monomials of the same degree. We say that

$$I_1 > I_2$$

provided:

(a): The first entry of I_1 is greater than the first entry of I_2;

or if these agree,

(b): in the first place where the entries differ, that of I_1 is greater than that of I_2.

Rephrasing this order in terms of excess reads, $I_1 > I_2$ if and only if

$$ex(I_1) > ex(I_2) \quad \text{in case (a)}$$

or

$$ex(I_{1,r}) > ex(I_{2,r}) \quad \text{in case (b)}$$

where the first $(r-1)$-entries agree, and the sequences $I_{1,r}$, $I_{2,r}$ are obtained from I_1, I_2 respectively by deleting the first $(r-1)$-entries from each. We call this ordering the *ordering of admissible monomials by excess.*

The divergence of our two ordering appears as soon as it can, in degree 12. The ordering by excess produces

$$\begin{array}{ccccccc} (12) > & (11,1) > & (10,2) > & (9,3) > & (9,2,1) > & (8,4) > & (8,3,1) \\ \xi_1^{12} & \xi_1^9\xi_2 & \xi_1^6\xi_2^2 & \xi_1^3\xi_2^3 & \xi_1^5\xi_3 & \xi_2^4 & \xi_1^2\xi_2\xi_3 \ , \end{array}$$

where the second row is $\xi(I')$ for $Sq(I)$ above it. The pair $\xi_1^5\xi_3$, ξ_2^4 is out of order in the lexicographic order from the right. That order would reverse these two elements with the rest of the list unchanged.

What is the incidence matrix for the ordering of admissible monomials by excess against the induced ordering of monomials in $\mathcal{A}^*$? It turns out that in each degree, this *matrix is also triangular.* We prove this using Milnor's theorem.

Proposition 1.2.6. If $A > B$ in the ordering by excess, then

$$\langle \xi(B'), Sq(A) \rangle = 0 \ .$$

Proof. Suppose $\ell n(A) = s$, $\ell n(B) = t$. Write

$$Sq(A) = Sq(a_1)Sq(a_2, \ldots, a_s) .$$

Now $\langle \xi_1^{a_1}, Sq(a_1)\rangle = 1$, and 0 is the result of any other monomial against $Sq(a_1)$. Hence in the coproduct for $\xi(B')$, we need only examine the contribution from terms of the form

$$(\xi_1 \otimes 1 + 1 \otimes \xi_1)^{b_1 - 2b_2}$$

and

$$(\xi_1^{2^{r-1}} \otimes \xi_{r-1})^{b_r - 2b_{r+1}} \; 2 \leq r \leq t .$$

Multiplying these together produces

$$\xi_1^{b_1} \otimes \xi((b_2, \ldots, b_t)') + \text{ terms of the form } \xi_1^u \otimes \gamma$$

with $u < b_1$. Thus if $a_1 > b_1$, the result follows. If $a_1 = b_1$, then $A > B$ holds as a consequence of the inequality

$$(a_2, \ldots, a_s, 0, \ldots) > (b_2, \ldots, b_t, 0, \ldots) ,$$

and an induction on length is possible. □

1.2.4. Negishi's theorem on left multiplication by Sq^{2^t}. The material in this section is due to A. Negishi [**87**]. Write $R(k)$ for the right ideal in the Steenrod algebra generated by certain dyadic squares,

$$R(k) = \left\{Sq^1, Sq^2, \ldots, Sq^{2^k}\right\}\mathcal{A} \quad \text{for } k \geq 0 ,$$
$$R(k) = 0 \quad \text{for } k < 0 .$$

Since the dyadic squares generate the Steenrod algebra, it follows that Sq^i is an element of $R(k)$ for all i satisfying

$$1 \leq i \leq 2^{k+1} - 1 .$$

Moreover, the diagonal map in $\mathcal{A}$ satisfies

$$\psi : R(k) \to R(k) \otimes \mathcal{A} + \mathcal{A} \otimes R(k) .$$

Let

$$L_k : \mathcal{A} \to \mathcal{A}$$

be left multiplication by Sq^{2^k},

$$L_k(a) = Sq^{2^k} a .$$

(1) *Negishi's Theorem. For each $k \geq 0$, L_k induces well-defined maps and an exact sequence,*

$$\mathcal{A}/R(k-3) \xrightarrow{{}_1L_k} \mathcal{A}/R(k-2) \xrightarrow{{}_2L_k} \mathcal{A}/R(k-1) .$$

Our proof of this theorem is based on Milnor's structure theorem. First of all, dualizing gives, (t denotes the annihilator)

$$(\mathcal{A}/R(k))^* = R(k)^t = Z/2[\xi_1^{2^{k+1}},\ \xi_2^{2^k}, \ldots, \xi_{k+1}^2, \xi_{k+2}, \ldots]$$

and $R(k)^t$ is a subalgebra of $\mathcal{A}^*$.

Next, we examine the map

$$L_k^* : \mathcal{A}^* \to \mathcal{A}^* .$$

In terms of the generic coproduct on an element α in $\mathcal{A}^*$

$$\varphi(\alpha) = \Sigma \alpha_i' \otimes \alpha_i'' ,$$

we have

$$L_k^*(\alpha) = \Sigma \langle Sq^{2^k},\ \alpha_i' \rangle \alpha_i''$$

and

$$\langle Sq^{2^k},\ \alpha_i' \rangle = 1 \Leftrightarrow \alpha_i' = \xi_1^{2^k} .$$

In particular, L_k^* is the zero map on $R(k)^t$. Moreover, for α in $\mathcal{A}^*$ and β in $R(k)^t$, we have

$$L_k^*(\alpha\beta) = (L_k^*(\alpha))\beta .$$

Similarly, we have

$$\begin{aligned} L_k^*(\alpha\xi_{k+1}) &= L_k^*(\alpha)\xi_{k+1} + \alpha\xi_k , \\ L_k^*(\alpha\xi_k) &= L_k^*(\alpha)\xi_k + L_{k-1}^*(\alpha)\xi_{k-1} , \\ L_k^*(\alpha\xi_k\xi_{k+1}) &= L_k^*(\alpha)\xi_k\xi_{k+1} + L_{k-1}^*(\alpha)\xi_{k-1}\xi_{k+1} + \alpha\xi_k^2 . \end{aligned}$$

In the case where α is a square, we have the squaring property,

$$L_k^*(\alpha) = (L_{k-1}^*\sqrt{\alpha})^2 .$$

To facilitate calculations with L_k^*, we observe that $R(k-1)^t$ is free over $R(k)^t$ and, as vector spaces, we have

$$R(k-1)^t \cong R(k)^t \otimes \Lambda(\xi_1^{2^k}, \xi_2^{2^{k-1}}, \ldots, \xi_{k+1}) .$$

It is now routine to establish that L_k^* determines well-defined maps in the dual sequence,

$$R(k-1)^t \xrightarrow{{}_2L_k^*} R(k-2)^t \xrightarrow{{}_1L_k^*} R(k-3)^t$$

and that the composition is 0.

To prove exactness, we first establish another detail concerning the map L_k^*. For this, we write $R(k-2)^t$ as

$$R(k-2)^t \cong Z/2[\xi_1^{2^{k-1}}, \xi_2^{2^{k-2}}, \ldots, \xi_k, \xi_{k+1}] \otimes Z/2[\xi_{k+2}, \ldots] .$$

Let A be the polynomial algebra displayed as the left factor.

Lemma 1.2.7. If homogeneous elements x, y in A satisfy the equations

$$L_k^* x = y\xi_{k+1} \ , \ L_k^* y = 0 \ ,$$

then $y = 0$ and $L_k^* x = 0$.

Proof. Introduce the subalgebra of A

$$B = Z/2[\xi_1^{2^k},\ \xi_2^{2^{k-1}}, \ldots, \xi_k^2, \xi_{k+1}]$$

and write

$$A = \Lambda(\xi_1^{2^{k-1}}, \ldots, \xi_k) \otimes B \ .$$

In terms of this decomposition, we have

$$x = \Sigma e_i \otimes b_i$$

where the set $\{e_i\}$ is a non-empty subset of the monomial basis for the exterior factor, and

$$L_k^*(x) = \Sigma((L_k^* e_i) b_i + e_i L_k^* b_i) \ .$$

Since $L_k^* e_i$ is in A only when it is 0, we must have

$$L_k^*(x) = \Sigma e_i L_k^* b_i \ .$$

Similarly, write $y = \Sigma f_i \otimes c_i$. Then the hypothesis on y yields that

$$L_k^*(c_i) = 0 \ .$$

To exploit this information, we write

$$B = B' \otimes \Lambda(\xi_{k+1})$$

where

$$B' = Z/2[\xi_1^{2^k}, \ldots, \xi_k^2, \xi_{k+1}^2] \ .$$

Write $c_i = c_{i,1} + c_{i,2}\xi_{k+1}$ with $c_{i,1}, c_{i,2}$ in B'. Then

$$0 = L_k^*(c_i) = L_k^*(c_{i,1}) + c_{i,2}\xi_k + L_k^*(c_{i,2}) \cdot \xi_{k+1} \ .$$

From the squaring property, it follows that $c_{i,2} = 0$. Similarly, write $b_i = b_{i,1} + b_{i,2}\xi_{k+1}$. Substitution in the equation of the lemma yields

$$\Sigma e_i \left(L_k^* b_{i,1} + b_{i,2}\xi_k + L_k^*(b_{i,2})\xi_{k+1}\right) = \Sigma f_i c_{i,1} \xi_{k+1} \ .$$

The summands not decorated with ξ_{k+1} add up to 0,

$$0 = \Sigma e_i (L_k^* b_{i,1} + b_{i,2}\xi_k) \ .$$

We now denote the indexing set for this sum as $I \cup J$, with empty intersection, where

$$i \in I \Leftrightarrow e_i \text{ does not have } \xi_k \text{ as a factor.}$$

For i in I we rewrite $e_i = e_{i,1}$ and for $i \in J$, we rewrite e_i as $e_{j,1}\xi_k$. Moreover, if $i \neq j$, then $e_{i,1} \neq e_{j,1}$ by considering degrees.

We may now write

$$0 = \sum_{i\in I} e_{i,1} L_k^* b_{i,1} + \left(\sum_{i\in I} e_{i,1} b_{i,2} + \sum_{j\in J} e_{j,1} L_k^* b_{j,1} \right) \xi_k + \left(\sum_{j\in J} e_{j,1} b_{j,2} \right) \xi_k^2 \, .$$

Thus $L_k^* b_{i,1} = 0$ for each i in I. Hence, in the coefficient of ξ_k^2, each $b_{j,2} = 0$. Thus, the coefficient of ξ_k is also 0. It follows that $b_{i,2} = 0$ for $i \in I$ and $L_k^* b_{j,1} = 0$ for $j \in J$. Hence we find that $L_k^* x = 0$. Thus $y = 0$ as well, completing the proof of the lemma.

Now the proof of Negishi's theorem goes by induction on k. The case $k = 0$ is a consequence of the Adem relations

$$Sq^1 Sq^1 = 0\, , \;\; Sq^1 Sq^{2n} = Sq^{2n+1}$$

and the Cartan basis of admissible monomials.

For the inductive step, we use the subalgebra A, which we now write as

$$(Z/2[\xi_1^{2^{k-2}}, \xi_2^{2^{k-3}}, \ldots, \xi_{k-1}, \xi_k, \xi_{k+1}])^2 \otimes \Lambda(\xi_k, \xi_{k+1}) \, .$$

Since ${}_1L_k^*$ annihilates $Z/2[\xi_{k+2}, \ldots]$, it is enough to prove that elements in the kernel of ${}_1L_k^*$ on A lie in the image of ${}_2L_k^*$. Now the image of ${}_1L_k^*$ lies in

$$Z/2[\xi_1^{2^{k-2}}, \xi_2^{2^{k-3}}, \ldots, \xi_{k-1}, \xi_k, \xi_{k+1}]$$

which we write as

$$(Z/2[\xi_1^{2^{k-3}}, \ldots, \xi_{k-2}, \xi_{k-1}, \xi_k, \xi_{k+1}])^2 \otimes \Lambda(\xi_{k-1}, \xi_k, \xi_{k+1}) \, .$$

Thus, for homogeneous x in A, we may write

$$x = x_0 + x_1 \xi_k + x_2 \xi_{k+1} + x_3 \xi_k \xi_{k+1}$$

for unique elements x_0, x_1, x_2, x_3 in the squared factor. We have (omitting the subscript 1)

$$\begin{aligned} L_k^*(x) = {} & (L_k^* x_0 + x_3 \xi_k^2) + (L_{k-1}^*(x_1)\xi_{k-1}) \\ & + (L_k^* x_1 + x_2)\xi_k + (L_k^* x_2 \xi_{k+1}) \\ & + (L_k^* x_3 \xi_k \xi_{k+1}) + L_{k-1}^* x_3 \xi_{k-1} \xi_{k+1} \, . \end{aligned}$$

Now, if $L_k^* x = 0$, then $L_k^* x_3 = 0$. By the squaring property, we have

$$L_{k-1}^* \sqrt{x_0} + \sqrt{x_3} \cdot \xi_k = 0$$

and $L_{k-1}^* \sqrt{x_3} = 0$. From the lemma, it follows that $x_3 = 0$ and $L_k^*(x_0) = 0$. By induction and the squaring property, we have

$$x_0 = {}_2L_k^* y_0 \, .$$

Furthermore, $x_2 = L_k^* x_1$. Hence

$$x = {}_2L_k^*(y_0 + x_1 \xi_{k+1}) \, ,$$

and the proof is complete. □

We apply Negishi's theorem in our proof of the Hopf invariant one result. The material in the next paragraph is used in this application. We refer to [**81**] and [**109**] for details.

The Steenrod algebra is equipped with the *canonical anti-automorphism*

$$\chi : \mathcal{A} \to \mathcal{A} .$$

This map is defined by induction on degree. Let $\chi(1) = 1$. If

$$\psi(a) = \Sigma a_i' \otimes a_i'' + a \otimes 1 + 1 \otimes a ,$$

then

$$\chi(a) = -a - \Sigma\chi(a_i')a_i'' .$$

The construction is motivated by the requirement that the composition

$$\mathcal{A} \xrightarrow{\psi} \mathcal{A} \otimes \mathcal{A} \xrightarrow{\chi \otimes Id} \mathcal{A} \otimes \mathcal{A} \xrightarrow{\varphi} \mathcal{A}$$

is 0. On products (anticipating odd primes) we have

$$\chi(ab) = (-1)^{|a|\,|b|}\chi(b)\chi(a)$$

and, as a consequence of the commutativity of ψ ,

$$\chi^2 = 1 .$$

In particular, from indecomposability we obtain

$$\chi(Sq^{2^k}) = Sq^{2^k} + \text{ decomposable terms.}$$

1.2.5. Steenrod operations which annihilate classes of a fixed dimension. We wish to identify elements of the Steenrod algebra which give 0 when applied to any cohomology class of a fixed dimension n. We call θ in $\mathcal{A}$ an *n-annihilator* provided

$$\theta(x) = 0 \quad \text{ if dimension } x \leq n .$$

Clearly, the set of n-annihilators forms a left ideal in $\mathcal{A}$, which we write as $Ann(n)$.

The main result of this section is *$Ann(n)$ has a basis comprised of all $Sq(I)$ where I is admissible and $ex(I) > n$.*

This result is included in Serre's Theorem quoted in Section 1.3.7. Here we give another proof based on the material in this section. Subsequent work in this book does not depend on the details that follow.

The strategy for the proof is to compare three ideals in $\mathcal{A}$. Let $V(n)$ be the $Z/2$ vector space

$$V(n) = \text{ span } \{Sq(I) | I \text{ admissible}, ex(I) > n\} .$$

Let $\widetilde{V(n)}$ be the left ideal in $\mathcal{A}$ generated by $V(n)$. From the zero property and the equation

$$\deg I + exI = 2i_1$$

where i_1 is the first entry of I, we obtain the inclusion

$$\widetilde{V(n)} \subset Ann(n) .$$

The third ideal is constructed from the observation that a sufficient condition for θ to be in $Ann(n)$ is that θ is the sum of monomials of the form

$$m_1 Sq(i) m_2 \quad \text{where } m_1 \text{ and } m_2 \text{ are monomials}$$

and

$$i + \deg(m_2) > n .$$

We write $B(n)$ for the left ideal in $\mathcal{A}$ spanned by these elements. Since $V(n) \subset B(n)$, we have the inclusion of left-ideals

$$\widetilde{V(n)} \subset B(n) \subset Ann(n) .$$

From the definition, it is straightforward to check that the application of an Adem relation to a monomial in $B(n)$ yields a sum of monomials in $B(n)$. Hence $B(n)$ is contained in $V(n)$ and

$$V(n) = \widetilde{V(n)} = B(n) .$$

We retain the notation $B(n)$ to denote the common object.

To complete the proof, we recall the space L_n and the evaluation map from Prop. 1.2.5, which we now write as

$$\lambda_n : \mathcal{A} \to H^*(L_n) ,$$
$$\lambda_n(\theta) = \theta(\sigma_n) .$$

Since $B(n) \subset \ker \lambda_n$, we have the extension

$$\bar{\lambda}_n : \mathcal{A}/B(n) \to H^*(L_n) .$$

Then our result follows from

Proposition 1.2.8. $\bar{\lambda}_n$ is a monomorphism.

To prove this, we make an induction on n. The case $n = 1$ is a direct verification. Thus $\mathcal{A}/B(1)$ is spanned by images of admissible monomials $Sq(I)$ with $ex(I) = 1$. Hence $I = (2^{i-1}, 2^{i-2}, \dots, 1)$ and $Sq(I)\sigma_1 = \sigma_1^{2^i} \neq 0$.

We now assume the truth of Prop. 1.2.8 for $k < n$. We have the following map of $\mathcal{A}$-modules,

$$\rho : H^*(L_n) \to H^*(L_k) \text{ for } k < n$$

obtained by formally setting $x_{k+1}, x_{k+2}, \ldots, x_n$ equal to 1 in any term from H^*L_n. Thus

$$\rho(\sigma_n) = \sigma_k$$

and the following diagram commutes:

$$\begin{array}{ccc} \mathcal{A}/B(n) & \xrightarrow{\bar{\lambda}_n} & H^*L_n \\ \downarrow & & \downarrow{\scriptstyle\rho} \\ \mathcal{A}/B(k) & \xrightarrow{\bar{\lambda}_k} & H^*L_k \, . \end{array}$$

From the filtration of $\mathcal{A}$ by the sub-modules $B(k)$, we have a sequence of inclusions

$$B(0)/B(n) \supset B(1)/B(n) \supset \cdots \supset B(n-1)/B(n) \, .$$

For $k < n$, it follows from the inductive hypothesis that the bottom horizontal arrow in the diagram below is a monomorphism:

$$\begin{array}{ccc} B(k-1)/B(n) & \longrightarrow & H^*(L_n) \\ \downarrow & & \downarrow \\ B(k-1)/B(k) & \longrightarrow & H^*(L_k) \, . \end{array}$$

Hence, the inductive step is complete when we show

$$B(n-1)/B(n) \to H^*(L_n)$$

is monic. This assertion will follow from the additive structure of $B(n-1)/B(n)$ described in

Lemma 1.2.9. The images of

$$\{Sq(2I)Sq(n) \mid I \text{ admissible, } ex(I) \leq n\}$$

give a basis for $B(n-1)/B(n)$, where $2I$ is obtained from I by doubling each entry.

Before proving this, we use it to complete the proof of Prop. 1.2.8. We have

$$Sq(2I)Sq(n)\sigma_n = Sq(2I)\sigma_n^2 = (Sq(I)\sigma_n)^2 \, .$$

It is tedious, but not difficult, to check that as I runs over admissible monomials of excess $\leq n$, the resulting set of elements $(Sq(I)\sigma_n)^2$ are linearly independent. First, do the case $ex(I) = n$, where $(i_2, \ldots, i_k)$ has excess $< n$. Then $i_1 = n + \sum_{j>2} i_j$ and

$$(Sq(I)\sigma_n)^2 = (Sq(i_2, \ldots, i_k)\sigma_n)^4 \, .$$

Thus as I runs over these sequences, the results form an independent set by invoking the induction hypothesis. A similar argument works if the excess of the subsequence equals n. Write

$$I = (I_1, I_2)$$

with I_1, the shortest sequence such that $exI_2 < n$. Then

$$(Sq(I)\sigma_n)^2 = (Sq(I_2)\sigma_n)^{2^{\ell n(I_1)}} .$$

Thus the set of admissible monomials I with $ex(I) \leq n$ is equivalent to the set

$$\{\ \{0\}, \{I_2\} \mid I = (I_1, I_2),\ exI_2 < n, I_1 \text{ of maximal length with respect to this decomposition}\ \}.$$

The inductive hypothesis implies that the set $Sq(J)\sigma_n$ is independent where J is taken from the set above. Their dyadic powers are independent in the polynomial algebra H^*L_n.

We turn to the proof of Prop. 1.2.9. First, the Adem relations yield

$$Sq(2I)Sq(n) \equiv Sq(n + \deg(I))Sq(I) \quad \mod B(n+1)$$

where I is admissible. Thus the elements in the displayed set are independent. To show that they span, we use Prop. 1.2.6 to observe that

$$(B(n-1)/B(n))^*$$

has a basis of the images of $\{\xi(J) \mid m(J) = n\}$. Next we map

$$(B(n-1)/B(n))^* \to \mathcal{A}^*$$

by using the coaction and evaluating against Sq^n,

$$\alpha \to \Sigma\alpha'_i \otimes \alpha''_i \to \Sigma\alpha'_i\langle\alpha''_i\, , Sq^n\rangle\ ,$$

where α'_2, α''_i are monomials in $\mathcal{A}^*$. Then

$$\langle\alpha''_i, Sq^n\rangle = 1 \Leftrightarrow \alpha''_i = \xi_1^n\ .$$

Inspection of the coproduct formula in $\mathcal{A}^*$ yields that our map is given by

$$\xi(n_1, \ldots, n_k) \to \xi(2n_2, 2n_3\, , \ldots, 2n_k)\ .$$

That is,

$$\xi_1^{n_1} \cdots \xi_k^{n_k} \to \xi_1^{2n_2} \cdots \xi_{k-1}^{2n_k}\ .$$

Now $n_2 + \cdots + n_k \leq n$ and any such sequence is possible. Thus comparing dimensions yields the spanning property and concludes proof of the lemma.

In subsequent discussions, we shall use the notation $B(n)$ to refer to the common object of this section. This is also the notation in the literature.

A corollary of the lemma is that $B(n)$ is generated as an $\mathcal{A}$-module by

$$\{Sq^{n+i} \mid i \geq 1\}\ .$$

This follows because each $B(k)/B(k+1)$ is a cyclic $\mathcal{A}$-module with generator the image of $Sq(k+1)$, and $B(k)$ is 0 in degrees $\leq k$. It is not difficult to extract a minimal set of generators from the set above. This is done in [**37**].

1.3. Cohomology operations and Eilenberg-Mac Lane spaces

In this section we recall the representation theorem for the cohomology functors. The principal consequence of this theorem is the identification of cohomology operations with elements in the cohomology of Eilenberg-Mac Lane spaces. In addition, we shall use the theorem to define maps up to homotopy and, in particular, to represent Steenrod operations as maps between Eilenberg-Mac Lane spaces.

Let π be an abelian group and $n \geq 1$. Let $K(\pi, n)$ denote an Eilenberg-Mac Lane space having the homotopy type of a CW complex. We write $[A, B]$ for the unpointed homotopy classes of maps from A to B.

1.3.1. Representation theorem for cohomology. Let (X, Y) be a CW pair. There is an isomorphism of abelian groups

$$H^n(X, Y; \pi) \cong [X/Y, K(\pi, n)]$$

given as follows. Let b_n in $H^n(K(\pi, n); \pi)$ be the fundamental class. Then for each cohomology class x in $H^n(X, Y; \pi)$ there exists a map

$$f : X/Y \to K(\pi, n)$$

unique up to homotopy, such that

$$f^*(b_n) = x .$$

We identify $H^n(X, Y; \pi)$ with $\bar{H}^n(X/Y; \pi)$ by collapsing Y to a 0-cell in X.

1.3.2. Pointed maps. Eventually, we work in a pointed category of pairs having the homotopy type of CW complexes. The map f provided by (3.1) is an unpointed map. But if $f : A \to B$ is a map of pointed CW complexes and B is path connected, then f is homotopic to a pointed map. Furthermore, if B is either simply connected or an H-space, a homotopy between pointed maps can be replaced by a basepoint preserving homotopy. Thus in (3.1) we can assume that

$$f : (X/Y, *) \to (K(\pi, n), *)$$

and the basepoints are 0-cells.

1.3.3. Suspension homomorphisms for cohomology. In cohomology theory there are two kinds of suspension homomorphisms in common use,

$$\underline{\text{Mayer-Vietoris suspension}}, \quad \Delta^* : \bar{H}^n(W) \to H^{n+1}(\Sigma W),$$

and

$$\underline{\text{Cohomology suspension}}, \quad \sigma^* : H^{n+1}(W) \to \bar{H}^n(\Omega W) .$$

The first is an isomorphism and the second is significant in applications of the representation theorem. In this section, we make the definitions and develop expressions for the two suspension operations in terms of the representation theorem.

Let Y be a pointed CW complex. Let ΣY denote the reduced suspension and let T_0Y, T_1Y be respectively the subspaces of ΣY obtained from $[0, \frac{1}{2}] \times Y$ and $[\frac{1}{2}, 1] \times Y$. We identify Y with $\{\frac{1}{2}\} \times Y$ in ΣY. The Mayer-Vietoris suspension is defined by the following isomorphism, induced by the inclusion of the pair

$$(T_0Y, Y) \subset (\Sigma Y, T_1Y)$$

and the connecting homomorphism for the pair (T_0Y, Y)

$$\Delta^* : \bar{H}^n(Y) \underset{\cong}{\overset{\delta}{\longrightarrow}} H^{n+1}(T_0Y, Y) \underset{\cong}{\longleftarrow} H^{n+1}(\Sigma Y, T_1Y) \underset{\cong}{\longrightarrow} H^{n+1}(\Sigma Y) .$$

Had we reversed the order of the cones, the result would have been the negative of our Δ^*.

If $f : Y \to Z$, then from the definition we have

$$\Delta^* \circ f^* = (\Sigma f)^* \circ \Delta^* .$$

Let (X, Y) be a pointed CW pair. We have a map

$$r : X/Y \to \Sigma Y$$

obtained by composing a homotopy equivalence

$$r_1 : X/Y \to X \cup T_1Y$$

with the map

$$r_2 : X \cup T_1Y \to \Sigma Y$$

obtained by collapsing X to the basepoint. Writing

$$p_0 : (X, Y) \to (X/Y, *)$$

for the quotient map, we have the commutative diagram

$$\begin{array}{ccc} \bar{H}^n(Y) & \xrightarrow{\delta} & H^{n+1}(X,Y) \\ \Delta^* \downarrow & & \uparrow p_0^* \\ H^{n+1}(\Sigma Y) & \xrightarrow[r^*]{} & H^{n+1}(X/Y) \end{array}$$

and we can identify δ with $r^* \circ \Delta^*$ via p_0^*. To see this, consider the following diagram:

$$\begin{array}{ccccc} (T_0X, Y) & \subset & (T_0X \cup T_1Y, T_1Y) & \supset & (\Sigma Y, *) \\ \cup & & \cup & \simeq & \uparrow r_2 \\ (X, Y) & \subset & (X \cup T_1Y, T_1Y) & \supset & (X \cup T_1Y, *) \\ \downarrow p_0 & & \downarrow & & p_0 \downarrow \uparrow r_1 \\ (X/Y, *) & = & (X/Y, *) & = & (X/Y, *) \end{array} .$$

The square in the upper right corner commutes up to homotopy by means of a homotopy contracting X to the basepoint in T_0X. We use the middle row to identify $H^{n+1}(X, Y)$ with $H^{n+1}(X \cup T_1Y)$; and we can calculate Δ^* using the top row. Since $r = r_2 \circ r_1$ and r_1p_0 is homotopic to the identity, the formula follows.

To define the cohomology suspension, let PW be the space of maps from I to the pointed space W such that 0 is mapped to the basepoint of W, and let $\Omega W \subset PW$ be the subspace of loops. We have a map of pairs

$$p : (PW, \Omega W) \to (W, *)$$

given by evaluation $p(w) = w(1)$. We also have the connecting homomorphism δ for the pair $(PW, \Omega W)$. Then the cohomology suspension is defined as

$$\sigma^* : H^*(W) \xrightarrow{p^*} H^*(PW, \Omega W) \underset{\cong}{\xrightarrow{\delta}} \bar{H}^*(\Omega W) \quad \sigma^* = \delta^{-1} \circ p^* .$$

If $f : Z \to W$, then

$$(\Omega f)^* \circ \sigma^* = \sigma^* \circ f^* .$$

Next, we turn to the expression of the two suspension homomorphisms in terms of the representation theorem. We shall employ the map

$$p_1 : \Sigma\Omega W \to W$$

given by

$$p_1(t, w) = w(t) .$$

The map p_1 is adjoint to the identity and is called the evaluation map. It satisfies

$$\Delta^* \circ \sigma^* = p_1^* .$$

Proof. Define $\lambda : T_0\Omega W \to PW$ by the formula

$$\lambda(t,w)(s) = w(2ts)\ .$$

We can extend λ by the identity map on $T_1\Omega W$ to

$$\bar{\lambda} : (\Sigma\Omega W, T_1\Omega W) \to (PW \cup T_1\Omega W, T_1\Omega W)\ .$$

Now the restriction of the evaluation map p_1 to $T_0\Omega W$ satisfies,

$$p_1 \mid T_0\Omega W \simeq p \circ \lambda$$

since $p \circ \lambda(t,w) = w(2t)$. Then the formula is a consequence of the commutative diagram

$$\begin{array}{ccccccc}
\bar{H}^n(\Omega W) & \xrightarrow{\ \delta\ } & H^{n+1}(T_0\Omega W, \Omega W) & \xleftarrow{\ \cong\ } & H^{n+1}(\Sigma\Omega W, T_1\Omega W) & \xrightarrow{\ \cong\ } & H^{n+1}(\Sigma\Omega W) \\
\| & & \uparrow{\scriptstyle \lambda^*} & & \uparrow{\scriptstyle \bar{\lambda}^*} & & \\
\bar{H}^n(\Omega W) & \xrightarrow{\ \delta\ } & H^{n+1}(PW, \Omega W) & \overset{\cong}{\leftarrow} & H^{n+1}(PW, \Omega W) \cup T_1\Omega W, T_1\Omega W) & & \uparrow{\scriptstyle p_1^*} \\
 & & \uparrow{\scriptstyle p^*} & & & & \\
 & & H^{n+1}(W) & & = & & H^{n+1}(W)
\end{array}$$

where the top row defines Δ^*. □

A useful corollary of this present argument concerns the connecting homomorphism for homotopy.

1.3.3.1. The map $p_\# \circ \partial^{-1}$

$$\pi_{n-1}(\Omega W) \underset{\cong}{\overset{\partial}{\longleftarrow}} \pi_n(PW, \Omega W) \xrightarrow{p_\#} \pi_n(W)$$

sends a class $[\alpha]$ to the homotopy class of the adjoint of α, since the adjoint of α is the composition $p_1 \circ \Sigma\alpha$, and $p_1 \mid T_0\Omega W$ is homotopic to $p \circ \lambda$.

The Hurewicz homomorphism, as described in [**117**] is compatible with the connecting homomorphism in homotopy and homology. Thus, we have

$$\sigma^* b_{n+1} = b_n$$

where b_m is the fundamental class, which we recall, corresponds to the identity under the following composition of isomorphisms

$$\begin{aligned}
H^m(K(\pi,m),\pi) &\cong Hom(H_n(K(\pi,m);Z),\pi) \text{ using universal coefficients} \\
&\cong Hom(\pi_m(K(\pi,m)),\pi) \text{ using Hurewicz,}
\end{aligned}$$

and $\pi = \pi_m K(\pi,m)$. We obtain the same result whether we use the suspension construction or the adjoint relation to identify coefficients with homotopy groups.

We can now describe Δ^* and σ^* in terms of the representation theorem.

Proposition 1.3.1. Let u be a class in $H^n(Y;\pi)$ and be represented by a map

$$f : Y \to \Omega K(\pi, n+1) .$$

Then $\Delta^*(u)$ is represented by the adjoint of f

$$f^\flat : \Sigma Y \xrightarrow{\Sigma f} \Sigma\Omega K(\pi, n+1) \xrightarrow{p_1} K(\pi, n+1)$$

and $\sigma^*(u)$ is represented by Ωf.

Proof.

$$\begin{aligned}\Delta^*(u) &= \Delta^* f^* b_n = (\Sigma f)^* \Delta^*(b_n) \\ &= (\Sigma f)^* \Delta^* \sigma^* b_{n+1} = (p_1 \circ \Sigma f)^* b_{n+1} ;\end{aligned}$$

and

$$\sigma^*(u) = \sigma^* f^*(b_n) = \Omega f^* \sigma^*(b_n) = \Omega f^*(b_{n-1}) .$$

□

Corollary 1.3.2. The value of the connecting homomorphism $\delta(u)$ is represented by $p_1 \circ \Sigma f \circ r \circ p_0$,

$$(X, Y) \xrightarrow{p_0} X/Y \xrightarrow{r} \Sigma Y \xrightarrow{f^\flat} K(\pi, n+1) .$$

1.3.4. Representation of cohomology operations as maps. An application of the representation theorem to the concept of a cohomology operation yields that the abelian group of operations of type (π, n, G, q) under addition of functions, is isomorphic to the abelian group $H^q(K(\pi, n); G)$ with the operation θ corresponding to the element $\theta(b_n)$. In turn, we can regard $\theta(b_n)$ as a map

$$h_\theta : K(\pi, n) \to K(G, q)$$

such that $h_\theta^*(b_n) = \theta(b_n)$. The map h_θ is defined up to homotopy.

By taking loops, we can define an operation

$$\Omega\theta \text{ of type } (\pi, n-1, G, q-1)$$

by the equations

$$\begin{aligned}(\Omega h_\theta)^*(b_{n-1}) &= (\Omega h_\theta)^* \sigma^*(b_n) = \sigma^* h_\theta^*(b_n) \\ &= \sigma^*(\theta b_n) .\end{aligned}$$

The pair of operations $\theta, \Omega\theta$ are compatible with the Mayer-Vietoris suspension,

$$\Delta^* \circ \Omega\theta = \theta \circ \Delta^* ,$$

since, in particular, if a cohomology class u in $H^{n-1}(Y)$ is represented by $f : Y \to \Omega K(\pi, n)$, then $\theta \circ \Delta^*(u)$ represented by the adjoint of $\Omega h_\theta \circ f$.

By Cor. 1.3.2, the pair of operations is also compatible with the connecting homomorphism,

$$\begin{array}{ccc} H^n(X,Y) & \xrightarrow{\theta} & H^q(X,Y) \\ \uparrow \delta & & \uparrow \delta \\ H^{n-1}(Y) & \xrightarrow[\Omega\theta]{} & H^{q-1}(Y) \end{array}$$

commutes. In particular, $\Omega\theta$ is additive.

Remark. $\Omega\theta$ is the unique map making the diagram above commute. For if θ' satisfies $\delta \circ \theta' = \theta \circ \delta$, then taking the case of δ for the pair

$$(PK(\pi, n), \Omega K(\pi, n)) \ ,$$

we have

$$\begin{aligned} \delta\theta'(b_{n-1}) &= \theta\delta(b_{n-1}) = \theta\delta\sigma^* b_n \\ &= \theta p^* b_n = p^*\theta b_n \ . \end{aligned}$$

Thus $\theta'(b_{n-1}) = \sigma^*\theta(b_n)$ and $\theta' = \Omega\theta$.

In view of the remark above, we can reformulate the notion of a stable cohomology operation in (1.1.2) as a sequence $\{\theta_n\}$ with $\Omega\theta_{n+1} = \theta_n$.

1.3.5. Loops, adjoints and suspensions. We summarize the compatibility of these concepts:

(a): The adjoint of the composition

$$X \xrightarrow{\epsilon} \Omega C_1 \xrightarrow{\Omega\theta} \Omega C_1$$

is θ composed with the adjoint of ϵ and $\Delta^* \circ \Omega\theta = \theta \circ \Delta^*$.

(b): The adjoint of the composition

$$\Sigma X \xrightarrow{\epsilon} C_0 \xrightarrow{\theta} C_1$$

is $\Omega\theta$ composed with the adjoint of ϵ and $\Omega\theta \circ \sigma^* = \sigma^* \circ \theta$.

We deliberately use the same symbol to represent an operation and a representative map.

1.3.6. Matrix representation of maps of generalized Eilenberg-Mac Lane spaces. Let C_0 and C_1 be finite products of Eilenberg-Mac Lane spaces,

$$C_0 = \prod_{j=1}^{a} K(Z/p, m_j) \stackrel{\text{def}}{=} K(m_1, \ldots, m_a) \ .$$

For indexing, we write b_j for the fundamental class of the j-th factor $K(Z/p, m_j)$ and likewise for

$$C_1 = \prod_{i=1}^{b} K(Z/p, n_i) \stackrel{\text{def}}{=} K(n_1, \ldots, n_b)$$

with fundamental classes b_i.

Given $f : C_0 \to C_1$, we represent f by a matrix with b rows and a columns

$$A_f = (\theta_{ij}) \quad 1 \leq i \leq b,\ 1 \leq j \leq a$$

where θ_{ij} is an operation of type

$$(Z/p, m_j, Z/p, n_i)$$

given by

$$f^* b_i = \sum_{j=1}^{a} \theta_{ij} b_j \ .$$

For example, take $p = 2$ and

$$C_0 = K(n),\ C_1 = K(n+1, n+2),\ C_2 = K(n+3, n+4)$$

with $f : C_0 \to C_1$ and $g : C_1 \to C_2$ given by $f^* b_{n+1} = Sq^1 b_n$, $f^* b_{n+2} = Sq^2 b_n$ and $g^* b_{n+3} = Sq^1 b_{n+2}$, $g^* b_{n+4} = Sq^3 b_{n+1} + Sq^2 b_{n+2}$. Then

$$A_f = \begin{pmatrix} Sq^1 \\ Sq^2 \end{pmatrix} ,$$

$$A_g = \begin{pmatrix} 0 & Sq^1 \\ Sq^3 & Sq^2 \end{pmatrix} .$$

The naturality property, written as a matrix product, reads

$$f^* \left(\sum_{i=1}^{b} \varphi_i b_{n_i} \right) = (\varphi_1, \ldots, \varphi_b)(\theta_{ij}) \begin{pmatrix} b_{m_1} \\ \vdots \\ b_{m_a} \end{pmatrix} .$$

We abbreviate this equation to

$$f^*(\bar{\varphi} \bar{b}_b) = \bar{\varphi} A_f \bar{b}_a \ ,$$

where

$$\bar{\varphi} = \ \text{row}\ (\varphi_1, \ldots, \varphi_b)\ ,$$
$$\bar{b}_a = \ \text{column}\ (b_{m_1}, \ldots, b_{m_a}) \text{ and likewise for } \bar{b}_b \ .$$

The notation is set up such that

$$A_{g \circ f} = A_g A_f \ .$$

Proof. $(g \circ f)^*(\bar{\varphi}\bar{b}_c) = \bar{\varphi}A_{g\circ f}\bar{b}_a$ and also equals

$$\begin{aligned} f^* \circ g^*(\bar{\varphi}\bar{b}_c) &= f^*(\bar{\varphi}A_g\bar{b}_b) \\ &= \bar{\psi}A_f\bar{b}_a \ , \end{aligned}$$

where $\bar{\psi} = \bar{\varphi}A_g$. □

In our example, we have

$$A_gA_f = \begin{pmatrix} 0 & Sq^1 \\ Sq^3 & Sq^2 \end{pmatrix} \begin{pmatrix} Sq^1 \\ Sq^2 \end{pmatrix} = \begin{pmatrix} Sq^3 \\ 0 \end{pmatrix}$$

by the Adem relations. This result agrees with A_{gf}.

In the sequel, we will with reluctance resist displaying the composition as

$$C_2 \overset{g}{\longleftarrow} C_1 \overset{f}{\longleftarrow} C_0 \ ,$$

but will remember to multiply representative matrices in the left to right order suggested by this display.

1.3.7. Serre's theorem on the cohomology of $K(A, n)$. For $A = Z/2$, Z, and $Z/2^f$, $f \geq 2$, we quote the results obtained by Serre in [**101**]. In each of the following, we describe the algebras over the Steenrod algebra.

(1) $H^*(K(Z/2, n); Z/2) = Z/2[Sq(I)b_n \mid I$ admissible, excess $(I) < n]$.

(2) Let u_n in $H^n(K(Z, n); Z/2)$ be the mod 2 reduction of the fundamental class,

$$H^*(K(Z, n); Z/2) = Z/2[Sq(I)u_n \mid I \text{ admissible, excess } (I) < n, I \text{ does not terminate with } 1].$$

(3) Let u_n in $H^n(K(Z/2^f, n); Z/2)$ be the mod 2 reduction of the fundamental class and let $v_{n+1} \epsilon H^{n+1}$ be the generator of

$$Z/2 \cong Ext(Z/2^f, Z/2) \cong H^{n+1}(K(Z/2^f, n); Z/2)$$

via the universal coefficient theorem. We have

$$H^*(K(Z/2^f, n); Z/2) = Z/2[Sq(I)u_n, Sq(J)v_{n+1} \mid I, J \text{ admissible, excess } (I) < n, \text{ excess } (J) < n+1, \text{ neither } I \text{ nor } J \text{ terminate with } 1].$$

In particular, $Sq^1u_n = Sq^1v_{n+1} = 0$ in (2) and (3). The relation between u_n and v_{n+1} in (3) is discussed in section 2 of Chapter 5.

1.4. Steenrod operations and maps between spheres

This section exhibits one of the principal methods in homotopy theory for gaining information about compositions of maps.

Let

$$f : S^{n+i-1} \to S^n$$

be a map of spheres, and let

$$T_f = S^n \cup_f e^{n+i}$$

be the mapping cone of f. We say Sq^i <u>detects</u> f if

$$Sq^i : H^n(T_f) \to H^{n+i}(T_f)$$

is an isomorphism. We have some examples at hand:

(a): $n \geq 1, f : S^n \to S^n$, degree $f = 2\times$ odd integer is detected by Sq^1;

(b): $n \geq 2$, $\eta_n : S^{n+1} \to S^n$, each Hopf map, defined using complex multiplication for $n = 2$ and suspensions for $n \geq 3$, is detected by Sq^2;

(c): $n \geq 4,;\nu_n : S^{n+3} \to S^n$, each Hopf map, defined using quaternionic multiplication for $n = 4$ and suspensions for $n \geq 5$, is detected by Sq^4;

(d): $n \geq 8$, $\sigma_n : S^{n+7} \to S^n$, each Hopf map, defined using Cayley multiplication for $n = 8$ and suspensions for $n \geq 9$, is detected by Sq^8.

Various compositions of these maps are defined, raising the question of which compositions are essential? We present a powerful argument, due to Adem [**5**], which addresses this question.

Adem's argument can be set up, starting with any composition,

$$X \xrightarrow{f} Y \xrightarrow{g} Z\ ,$$

and works by deriving a contradiction from the supposition that the composition is null-homotopic. On this supposition, we have a commutative diagram

$$\begin{array}{ccc} X & \xrightarrow{f} & Y \\ \downarrow & & \downarrow g \\ CX & \longrightarrow & Z \end{array}$$

where CX is a cone on X. (Say $CX = I \times X/\{0\} \times X$ for the moment; we shall be more precise in Chapter 3.) By naturality, we receive a map

$$\underset{\sim}{f} : \Sigma X \to T_g = \text{ mapping cone of } g,$$

such that if

$$q : T_g \to \Sigma Y$$

is obtained by pinching Z to a point in T_g, then

$$q \circ \underset{\sim}{f} = \Sigma f \text{ up to homotopy.}$$

The map $\underset{\sim}{f}$ is called a coextension of f and is discussed further in Chapter 3.

We can now form the mapping cone of $\underset{\sim}{f}$,

$$T_{\underset{\sim}{f}} = Z \cup_g CY \cup_{\underset{\sim}{f}} C\Sigma X \ .$$

Furthermore, pinching Z to a point in $T_{\underset{\sim}{f}}$ yields a map

$$\hat{q} : T_{\underset{\sim}{f}} \to T_{\Sigma f} = \text{ mapping cone of } \Sigma f$$

which extends q. Namely, the diagram

$$\begin{array}{ccc} T_g & \xrightarrow{q} & \Sigma Y \\ {\scriptstyle j}\downarrow & & \downarrow \\ T_{\underset{\sim}{f}} & \xrightarrow{\hat{q}} & T_{\Sigma f} \end{array}$$

commutes up to homotopy, where the vertical maps are inclusions.

We shall use the term "Adem argument" to signify an argument that $T_{\underset{\sim}{f}}$ is an impossible space; hence, that $g \circ f$ is essential.

In the following examples, we have the additional information,

$$j^* \text{ is an epimorphism and } \hat{q}^* \text{ is a monomorphism.}$$

In these examples, the impossibility of $T_{\underset{\sim}{f}}$ will follow using naturality and the Adem relations.

We begin with the composition $\eta_2 \circ \eta_3$. Assuming this composition is 0, we obtain the diagram

$$\begin{array}{ccc}
S^2 \cup_{\eta_2} e^4 & \longrightarrow & S^4 \\
{\scriptstyle j}\downarrow & & \downarrow \\
S^2 \cup_{\eta_2} e^4 \cup_{\underset{\sim}{\eta_3}} e^6 & \underset{\hat{q}}{\longrightarrow} & S^4 \cup_{\eta_4} e^6
\end{array}$$

with j^* epic and $\hat{q}^*$ monic for dimensional reasons. Since Sq^2 detects each η, it follows that

$$Sq^2 Sq^2 \neq 0 \text{ in } T_{\underset{\sim}{\eta_3}}, \text{ in the lower left corner.}$$

But, $Sq^2 Sq^2 = Sq^3 Sq^1$ and the latter composition of Steenrod operations is 0 in the cohomology of $T_{\underset{\sim}{\eta}_3}$, for dimensional reasons. Hence $\eta_2 \circ \eta_3 \neq 0$ and likewise for all its suspensions.

Exercise 1.4.1. Show that the composition

$$S^3 \xrightarrow{[2]} S^3 \xrightarrow{\eta_2} S^2$$

is essential. One can use the Adem relation

$$Sq^1 Sq^2 = Sq^3 .$$

Remark. We have the diagram for the Adem argument provided that $g \circ f$ is null-homotopic. One can ask whether there is a converse. Namely, given a space T fitting in a homotopy commutative diagram

$$\begin{array}{ccccc}
 & & \Sigma X & =\!=\!= & \Sigma X \\
 & & \downarrow & & \downarrow{\scriptstyle \Sigma f} \\
Z & \longrightarrow & T_g & \longrightarrow & \Sigma Y \\
\| & & \downarrow & & \downarrow \\
Z & \longrightarrow & T & \longrightarrow & T_{\Sigma f} ,
\end{array}$$

does it follow that $g \circ f$ is null-homotopic? The answer is, not necessarily. One can construct the diagram

$$\begin{array}{ccccc} & & S^4 & = & S^4 \\ & & \downarrow & & \downarrow \eta_3 \\ S^2 & \longrightarrow & S^2 \cup_{[2]} e^3 & \longrightarrow & S^3 \\ \| & & \downarrow & & \downarrow \\ S^2 & \longrightarrow & S^2 \cup_{[2]} e^3 \cup e^5 & \longrightarrow & S^3 \cup_{\eta_3} e^5 , \end{array}$$

but $S^3 \xrightarrow{\eta} S^2 \xrightarrow{[2]} S^2$ is essential. To supply details is a good exercise, and we shall return to this example later in subsection 5.2.2. It is easy to see that $\Sigma g \circ \Sigma f$ is null-homotopic, in general, given T as above.

By specializing in the general formula, check the following Adem relations:

$$\begin{aligned} &Sq^1 Sq^4 + Sq^2 Sq^3 + Sq^4 Sq^1 = 0 , \\ &Sq^1 Sq^8 + Sq^8 Sq^1 + Sq^2 Sq^7 = 0 , \\ &Sq^2 Sq^8 + Sq^8 Sq^2 + Sq^4 Sq^6 + Sq^9 Sq^1 = 0 , \\ &Sq^4 Sq^4 + Sq^7 Sq^1 + Sq^6 Sq^2 = 0 , \\ &Sq^8 Sq^8 + Sq^{15} Sq^1 + Sq^{14} Sq^2 + Sq^{12} Sq^4 = 0 . \end{aligned}$$

Prove that each of the following compositions is essential:

$$\begin{array}{cl} ([2]) \circ \nu_n \text{ and } \nu_n \circ ([2]) & n \geq 4 , \\ ([2]) \circ \sigma_n \text{ and } \sigma_n \circ ([2]) & n \geq 8 , \\ \eta_n \circ \sigma_{n+1} & n \geq 7 , \\ \sigma_n \circ \eta_{n+7} & n \geq 8 , \\ \nu_n \circ \nu_{n+3} & n \geq 4 , \\ \sigma_n \circ \sigma_{n+7} & n \geq 8 . \end{array}$$

Remark. One should not assume compositions are equal if they happen to have the same source and target.

Exercise 1.4.2. Use the Adem relation

$$Sq^2 Sq^4 = Sq^6 + Sq^5 Sq^1$$

to prove that $\nu_n \circ \eta_{n+3}$ is essential for $n = 4, 5$. Observe that the presence of Sq^6 prevents exploitation of the Adem argument for $n \geq 6$.

Exercise 1.4.3. Use the Adem relation

$$Sq^4 Sq^8 = Sq^{12} + Sq^{11} Sq^1 + Sq^{10} Sq^2$$

to prove that $\sigma_n \circ \nu_{n+7}$ is essential for $n = 8, 9, 10, 11$; and we are "blocked" by Sq^{12} for $n \geq 12$.

Remark. The Adem argument gives no information for $\eta_n \circ \nu_{n+1}, n \geq 3$. It turns out that the composition is essential for $n = 3, 4$, but inessential for $n \geq 5$. See the remark at the end of (7.2.3).

1.5. Steenrod operations for odd primes

There are stable operations, called reduced powers of type $(Z/p, n, Z/p, n + 2i(p-1))$ for p an odd prime and written

$$\mathcal{P}^i : H^n(X, Y; Z/p) \to H^{n+2i(p-1)}(X, Y; Z/p) .$$

In addition, there is the Bockstein homomorphism

$$\beta : H^n(X, Y; Z/p) \to H^{n+1}(X, Y; Z/p)$$

obtained from the coefficient sequence

$$\begin{gathered} 0 \to Z/p \to Z/p^2 \to Z/p \to 0 \\ [1] \to [p] \ . \end{gathered}$$

The Bockstein operation anti-commutes with the connecting homomorphism for the general reason pointed out in [**23**]. We write

$$\mathcal{B} = (-1)^n \beta$$

for β defined on $H^n(X, Y; Z/p)$. Then the signed Bockstein $\mathcal{B}$ is a stable operation, and the sign is determined by the dimension of the source. Alternatively, the signed Bockstein is the connecting homomorphism induced by

$$\begin{gathered} 0 \to Z/p \to Z/p^2 \to Z/p \to 0 \\ [1] \to [(-1)^n p] \ . \end{gathered}$$

1.5.1. Properties of the reduced powers.

(1) $\mathcal{P}^0$ is the identity.

(2) If $n = 2i$, then $\mathcal{P}^i x = x^p$ for any cohomology class x of dimension n.

(3) Zero property. If $n < 2i$, then $\mathcal{P}^i$ is the zero map on the cohomology of dimension n. Since

$$\beta x^p = 0$$

we also have that $\beta \mathcal{P}^i$ is the zero map on the cohomology of dimension $n \leq 2i$.

(4) Cartan Formula.

$$\mathcal{P}^i(x\cdot y)=\sum_{j+k=i}\mathcal{P}^j x\cdot\mathcal{P}^k y\,,$$

and, for the unsigned Bockstein,

$$\beta(x\cdot y)=\beta x\cdot y+(-1)^{|x|}x\cdot\beta y\,.$$

(5) Adem relations. If $a<pb$,

$$\mathcal{P}^a\mathcal{P}^b=\sum(-1)^{a+t}\binom{(p-1)(b-t)-1}{a-pt}\mathcal{P}^{a+b-t}\mathcal{P}^t$$

with non-zero summands only for integers t satisfying

$$0\le t\le a/p\,.$$

If $a\le pb$, then

$$\mathcal{P}^a\beta\mathcal{P}^b=\sum(-1)^{a+t}\binom{(p-1)(b-t)}{a-pt}\beta\mathcal{P}^{a+b-t}\mathcal{P}^t$$
$$+\sum(-1)^{a+t-1}\binom{(p-1)(b-t)-1}{a-pt-1}\mathcal{P}^{a+b-t}\beta\mathcal{P}^t$$

and $\beta^2=0$.

Compositions of reduced powers and Bocksteins are handled with the following notation. Let

$$I=(\varepsilon_0,s_1,\varepsilon_1,\ldots,s_k,\varepsilon_k,0,\ldots)$$

be a sequence with each s_i a positive integer and $\varepsilon_i=0$ or 1. Write

$$\mathcal{P}(I)=\beta^{\varepsilon_0}\mathcal{P}^{s_1}\beta^{\varepsilon_1}\cdots\mathcal{P}^{s_k}\beta^{\varepsilon_k}$$

where $\beta^0=1$ and $\beta^1=\beta$. The degree of $\mathcal{P}(I)$ is given by

$$deg\ \mathcal{P}(I)=\sum_{j\ge0}\varepsilon_j+\sum_{i\ge1}2s_i(p-1)\,.$$

We say that I is admissible provided

$$s_i\ge ps_{i+1}+\varepsilon_i$$

and call $\mathcal{P}(I)$ admissible if I is. By fiat, $\mathcal{P}^0$ is admissible. The excess of an admissible monomial I, written $ex(I)$, is given by the formula

$$ex(I)=2\sum_{j\ge1}(s_j-ps_{j+1})+\varepsilon_0-\sum_{j\ge1}\varepsilon_j\,.$$

Note the special role for ε_0 in this formula. Moreover, we have

$$ex(I)+deg(I)=2(\varepsilon_0+ps_1)\,.$$

The Steenrod algebra $\mathcal{A}$ is constructed from the Bockstein and the reduced powers in the same manner as in section 2.

1.5.2. Cartan's Theorem.

(a): *In grade n, the Steenrod algebra has a basis over Z/p consisting of all admissible monomials of degree n.*

(b): *As an algebra over the Steenrod algebra, the mod p cohomology of $K(Z/p, n)$ is the free commutative algebra generated by the elements*

$$\beta^{\varepsilon}\mathcal{P}(I)b_n \ , \ \varepsilon = 0, 1$$

where I is admissible and $\varepsilon_0 = 0$, and $ex(I) < n$.

(c): *The left ideal $B(n)$ in the Steenrod algebra consisting of n-annihilators is spanned by all admissible monomials of excess $> n$.*

Next, we state Milnor's structure theorem for the dual Hopf algebra $\mathcal{A}^*$. We first define elements in the basis dual to the Cartan basis,

$$\xi_k \text{ is dual to } \mathcal{P}^{p^{k-1}}\mathcal{P}^{p^{k-2}} \cdots \mathcal{P}^1, \ k \geq 1 \ ,$$

$$\tau_k \text{ is dual to } \mathcal{P}^{p^{k-1}} \cdots \mathcal{P}^1\beta \ , \ k \geq 0 \ .$$

Degrees are given by

$$deg\ \xi_k = 2p^k - 2 \ , \ deg\ \tau_k = 2p^k - 1 \ .$$

Theorem 1.5.1 (Milnor's theorem)**.** The Steenrod algebra is a Hopf algebra. The dual Hopf algebra is a free commutative algebra over Z/p given by

$$Z/p[\xi_1, \xi_2, \dots] \otimes \Lambda(\tau_0, \tau_1, \dots) \ .$$

The coalgebra structure is given by

$$\xi_k \to \sum_{i=0}^{k} \xi_{k-i}^{p^i} \otimes \xi_i \ ,$$

$$\tau_k \to \tau_k \otimes 1 + 1 \otimes \tau_k + \sum_{i=0}^{k-1} \xi_{k-i}^{p^i} \otimes \tau_k \ .$$

We refer to [**20**], [**81**], [**109**], [**100**] for the facts presented here. Readers of [**100**] should note a usage for "excess" which differs from ours.

1.5.3. Sundry facts. We collect some further facts used in our calculations. The verifications are left as exercises.

(a): If u is a cohomology class of dimension 2, then

$$\mathcal{P}^i u^k = \binom{k}{i} u^{k+(p-1)i} \ .$$

(b): If $a < p$, the Adem relations simplify to

$$\mathcal{P}^a\mathcal{P}^b = \binom{a+b}{a}\mathcal{P}^{a+b},$$

$$\mathcal{P}^a\beta\mathcal{P}^b = \binom{a+b-1}{a}\beta\mathcal{P}^{a+b} + \binom{a+b-1}{b}\mathcal{P}^{a+b}\beta .$$

(c): For x of even degree, $\mathcal{P}^1$ is a derivation

$$\mathcal{P}^1 x^k = kx^{k-1}\mathcal{P}^1 x .$$

(d): (For use in Chapter V, section 2.) For p odd, there is no sequence I of degree $2n(p-1)+1$, $\varepsilon_0 = 0$ and $ex(I) = 2n$. For $p = 2$, the only admissible sequence of degree $2n+1$ with i_1 even and satisfying $2n \geq ex(I) \geq 2n-1$, is $I = (2n, 1)$.

(e): $\mathcal{P}^1$ is non-zero on $H^4(HP^2; Z/3)$. Suggestion, use HP^3.

(f): Use $\mathcal{P}^1$ for $p = 3$ to distinguish between

$$CP^6/CP^3 \text{ and } S^8 \vee \Sigma^8 CP^2 .$$

1.5.4. Left multiplication by $\mathcal{P}^{p^k}$. The material in this section is parallel to that discussed in subsection 1.2.4. It is due to Toda and Mukohda [**86**] . We follow Shimada and Yamanoshita in its use for the mod p Hopf invariant one theorem.

Consider the following right ideal in the Steenrod algebra

$$R(k) = \{\mathcal{P}^{p^k}, \mathcal{P}^{p^{k-1}}, \ldots, \mathcal{P}^1, \beta\}\mathcal{A} \qquad k \geq 0 ,$$
$$R(-1) = \{\beta\}\mathcal{A} \qquad k = -1 ,$$
$$R(k) = 0 \qquad k < -1 .$$

For $k \geq -1$, define

$$L_k : \mathcal{A} \to \mathcal{A}/R(k-1)$$

by $L_k(a) = [\mathcal{P}^{p^k} a]$ for $k \geq 0$,

$$L_{-1}(a) = \beta a .$$

It is easy to establish that ker $L_{-1} = R(-1)$. From the paper by Mukohda cited above, we quote

Theorem 1.5.2 (Mukohda's Theorem)**.** For $k \geq 1$, the kernel of L_k is given by

(a) $\quad R(k-2) + \mathcal{P}^{2p^{k-1}} \cdot \mathcal{A} + (2\mathcal{P}^{p^k+p^{k-1}} - \mathcal{P}^{p^k}\mathcal{P}^{p^{k-1}})\mathcal{A} + \mathcal{P}^{p^k(p-1)} \cdot \mathcal{A}$.

The kernel of L_0 is given by

(b) $\quad (2\beta\mathcal{P}^1 - \mathcal{P}^1\beta) \cdot \mathcal{A} + \mathcal{P}^{p-1}\mathcal{A}$.

Remark. I have not worked out a proof employing Milnor's theorem, as in the proof for Negishi's theorem. Mukohda argues with the Milnor basis for the Steenrod algebra.

In our proof of the mod p Hopf invariant one result, we use the fact

$$\mathcal{X}(\mathcal{P}^{p^k(p-1)}) \equiv \mathcal{P}^{p^k(p-1)} \text{ modulo } \{\mathcal{P}^1, \dots, \mathcal{P}^{p^{k-1}}\}$$

where curly brackets mean the two-sided ideal in $\mathcal{A}$.

1.5.5. Steenrod operations and maps between spheres. Iterated suspensions of HP^2 produce complexes

$$S^n \cup e^{n+4} \ , \ n \geq 4$$

supporting a non-zero $\mathcal{P}^1$ in mod 3 cohomology.

It follows that there is a map from S^{n+3} to S^n which is detected by $\mathcal{P}^1$ for $p = 3$. We can use localization theory to find more. Recall that

$$H^*(CP^n; Z/p) = Z/p[y]/y^{n+1} \ , \ \text{degree } y = 2 \ .$$

Then for $n = p$,

$$\mathcal{P}^1(y) = y^p \neq 0 \ .$$

There is a splitting of the p-localizations

$$(\Sigma CP^n)_{(p)} \cong (S^3 \cup e^{2p+1} \vee S^5 \vee \cdots \vee S^{2p-1})_{(p)} \ ,$$

and we refer to [**24**] for the method. It follows that there is a complex of the form

$$S^3 \cup e^{2p+1}$$

which supports a non-zero $\mathcal{P}^1$ in mod p cohomology, and by stability, likewise for the iterated suspensions.

Definition 1.5.3. $\alpha_1(n) : S^{n+2p-2} \to S^n$ for $n \geq 3$ denotes a map detected by $\mathcal{P}^1$, for $n = 3$, and the $(n-3)$ fold suspension of $\alpha_1(3)$ for $n > 3$.

Remark. It turns out that all such maps are multiples of a common one, since the p-component of $\pi_{2p}S^3$ is Z/p.

Proposition 1.5.4. The composition $\alpha_1(n) \circ \alpha_1(n + 2p - 3)$ is essential for $n = 3, 4$.

Proof. We use the Adem argument and the relation

$$\mathcal{P}^1\mathcal{P}^1 = 2\mathcal{P}^2 \ ,$$

for the case $n = 3$. If the composition is null-homotopic, we have a diagram of complexes and maps,

$$\begin{array}{lcl} T_{\alpha_1(n)}: & S^n \cup_{\alpha_1(n)} e^{n+2p-2} & \\ & \downarrow & \text{inclusion} \\ T_{\underset{\sim}{\alpha_1}}(n+2p-3): & S^n \cup_{\alpha_1(n)} e^{n+2p-2} \cup_{\underset{\sim}{\alpha_1}(n+2p-3)} e^{n+4p-4} & \\ & \downarrow & \text{pinch } S^n \\ T_{\Sigma\alpha_1(n+2p-3)}: & S^{n+2p-2} \cup_{\alpha_1(n+2p-2)} e^{n+4p-4} & \end{array}$$

with $\mathcal{P}^1$ non-zero in the mod p cohomology of the top and bottom complexes. But if $n = 3$, then $\mathcal{P}^2$ is 0 in the middle complex by the zero property.

A similar, but more delicate argument works for $n = 4$. Alternatively for $n = 4$, we can observe that the suspension map

$$E : S^3 \to \Omega S^4$$

induces a monomorphism on homotopy, because S^3 is an H-space. Since $\alpha_1(4) \circ \alpha_1(2p+1)$ is the suspension of $\alpha_1(3) \circ \alpha_1(2p)$, the case $n = 4$ follows from $n = 3$. □

Remark. It turns out that $\alpha_1(n) \circ \alpha_1(n+2p-3) = 0$ for $n \geq 5$. The method of connective covers provides a means to prove that the p-component of $\pi_{4p-1}S^5$ is 0.

The information provided by the remark above, that certain compositions are 0, can be used to construct new maps. These are called "secondary compositions", and we introduce them by example. Consider the following sequence of compositions, in which each pair of compositions is null-homotopic:

$$S^{6p-4} \overset{\alpha_1(4p-1)}{\longrightarrow} S^{4p-1} \overset{\alpha_1(2p+2)}{\longrightarrow} S^{2p+2} \overset{\alpha_1(5)}{\longrightarrow} S^5 .$$

We have an "extension"

$$\underline{\alpha_1(5)} : T_{\alpha_1(2p+2)} \longrightarrow S^5$$

where T_f denotes a mapping cone of f. We also have a "coextension"

$$\underset{\sim}{\alpha_1}(4p-1) : S^{6p-3} \longrightarrow T_{\alpha_1(2p+2)}$$

as discussed in Section 4. Since both these constructions depend on the homotopy chosen for the null-homotopy, our discussion here is rough. The detailed development is carried out in Chapter 3. Here the results are understood to hold for any composition produced by the construction.

We can form the "secondary composition" by composing the two maps,

$$\underline{\alpha_1(5)} \circ \underset{\sim}{\alpha_1}(4p-1) : S^{6p-3} \to S^5 ;$$

and ask whether it is null? The Adem argument can be applied provided we have means for "detecting" the two factors. Now all the α_1's are detected by $\mathcal{P}^1$. From the Adem relation $\mathcal{P}^1\mathcal{P}^1 = 2\mathcal{P}^2$, it follows that $\mathcal{P}^2$ is an isomorphism detecting $\underline{\alpha_1(5)}$,

$$\mathcal{P}^2 : H^5(T_{\underline{\alpha_1(5)}}) \to H^{4p+1}(T_{\underline{\alpha_1(5)}})$$

for the mod p cohomology of the mapping cone of $\underline{\alpha_1(5)}$. Indeed, we obtained this information in the proof of 1.5.4, where it "blocked" further argument.

For the other factor, recall that composition of the coextension $\underset{\sim}{\alpha_1}(4p-1)$ with the pinch map of the bottom cell gives the composition

$$\alpha_1(4p) : S^{6p-3} \to T_{\alpha_1(2p+2)} \to S^{4p} \ .$$

Thus we have an isomorphism

$$\mathcal{P}^1 : H^{4p}(T_{\underset{\sim}{\alpha_1}(4p-1)}) \to H^{6p-4}(T_{\underset{\sim}{\alpha_1}(4p-1)})$$

for the mod p cohomology of the mapping cone of $\underset{\sim}{\alpha_1}(4p-1)$.

We are now ready to use the Adem argument. Thus, if we assume that

$$\underline{\alpha_1(5)} \circ \underset{\sim}{\alpha_1}(4p-1) = 0 \ ,$$

then we have a map, the coextension of $\underset{\sim}{\alpha_1}(4p-1)$,

$$A : S^{6p-4} \to T_{\underline{\alpha_1(5)}}$$

such that composing A with the pinch map of the bottom cell in the mapping cone of $\underline{\alpha_1(5)}$ yields the suspension of $\underset{\sim}{\alpha_1}(4p-1)$,

$$\Sigma\underset{\sim}{\alpha_1}(4p-1) : S^{6p-4} \to \Sigma T_{\alpha_1(2p+2)} \ .$$

As in previous uses of the Adem argument, we display the key spaces and maps, now keeping track of dimensions of the cells:

$$\begin{array}{rcccccl} T_{\underline{\alpha_1(5)}} & : & (5) & (2p+3) & (4p+1) & & , \\ T_A & : & (5) & (2p+3) & (4p+1) & (6p-3) & , \\ T_{\Sigma\underset{\sim}{\alpha_1}(4p-1)} & : & & (2p+3) & (4p+1) & (6p-3) & . \end{array}$$

We have established that $\mathcal{P}^2$ is non-zero in the top row. Furthermore, stability of $\mathcal{P}^1$ and the calculation of $\mathcal{P}^1$ for the cohomology of the mapping cone of $\underset{\sim}{\alpha_1}(4p-1)$, yields that the composition

$$\mathcal{P}^1\mathcal{P}^2 : H^5(T_A) \to H^{6p-3}(T_A)$$

is non-zero in the mod p cohomology for the mapping cone of A. Now if $p = 3$, this composition is 0, hence for $p = 3$ we have constructed an essential map (we use its standard name)

$$\beta_1 : S^{15} \to S^5 ;$$

and if $p \geq 5$, the relation reads $\mathcal{P}^1\mathcal{P}^2 = 3\mathcal{P}^3$, which is 0 by the zero property.

Exercise 1.5.5. For $p = 3$, prove that the above argument works to produce essential maps

$$\underline{\alpha_1}(n) \circ \underset{\sim}{\alpha_1}(n+6) : S^{n+10} \to S^n$$

starting with $\alpha_1(n) : S^{n+3} \to S^n$ and the information that compositions of the α_1's are 0.

Remark. For $p \geq 5$, our argument also works for $n = 6$, because S^5 is a p-local H-space. It turns out that the double suspension of $\underline{\alpha_1}(5) \circ \underset{\sim}{\alpha_1}(4p-1)$ is 0.

Chapter 2

Segue to Secondary Operations

This chapter does not contain any formal mathematics. Its aim is to compensate for the dry, nit-picking nature of the next chapter. In the preface, we suggested that secondary cohomology operations pick up, after primary operations fail to settle an issue. A typical way for this to happen is that a primary operation gives 0 in the cohomology of a space X. The representation theorem interprets this situation as a null-composition

$$X \xrightarrow{\epsilon} C_0 \xrightarrow{\theta} C_1 .$$

Relations among primary operations supply more null-compositions,

$$C_0 \xrightarrow{\theta} C_1 \xrightarrow{\varphi} C_2 .$$

Thus "secondary compositions" can be constructed which may give more information about X.

To help fix ideas, and serve as a running example, we discuss how a space X of the homotopy type

$$X = S^n \cup e^{n+3}$$

might be distinguished from the bouquet of spheres,

$$S^n \vee S^{n+3} .$$

Our answer, even in general, comes in two parts. The first part is to construct operations analogous to the primary operations. The second part concerns methods for making calculations with the new operations.

Take $p = 2$. We record the information that Sq^1 and Sq^2 (and hence all Steenrod operations) give 0 in the cohomology of X. The following composition is null,

$$X \xrightarrow{\epsilon} K(n) \xrightarrow{\theta} K(n+1) \times K(n+2), \quad \theta = \begin{pmatrix} Sq^1 \\ Sq^2 \end{pmatrix},$$

where $K(m) = K(Z/2, m)$ $\epsilon^* b_n \neq 0$ and the conventions introduced in (1.3.6) are employed. We have the Adem relation

$$Sq^2 Sq^2 + Sq^3 Sq^1 = 0$$

which we represent as a null composition $\varphi\theta$, where,

$$\varphi : K(n+1) \times K(n+2) \to K(n+4), \ \varphi = (Sq^3, Sq^1).$$

Let's first concentrate on the part that does not involve X. Since $\varphi \circ \theta$ is null-homotopic, we have the commutative diagram,

$$\begin{array}{ccc} K(n) & \longrightarrow & PK(n+4) = \text{ space of paths} \\ \theta \downarrow & & \downarrow \\ K(n+1, n+2) & \xrightarrow[\varphi]{} & K(n+4). \end{array}$$

We can take fibres of the vertical maps to produce the diagram

$$\begin{array}{ccc} F & \xrightarrow{\tilde{\varphi}} & \Omega K(n+4) \\ \downarrow & & \downarrow \\ K(n) & \longrightarrow & PK(n+4) \end{array}$$

and we call the map $\tilde{\varphi}$ a "colifting". As was the case in the previous chapter, these constructions depend on choices of homotopies, among other things, so this discussion will be rough.

When we insert X into the diagram above with the map ϵ,

$$\begin{array}{ccccc} X & \xrightarrow{\bar{\epsilon}} & F & \xrightarrow{\tilde{\varphi}} & \Omega K(n+4) = K(n+3) \\ \| & & \downarrow & & \\ X & \xrightarrow[\epsilon]{} & K(n) & & \end{array}$$

we have a "lifting" $\bar{\epsilon}$ of ϵ, since the condition $\theta \circ \epsilon$ is null-homotopic is exactly what is required.

We can use the map $\tilde{\varphi}$ to construct a certain function with domain

$$S(X) = \{x \in H^n(X; Z/2) \mid Sq^1 x = 0 = Sq^2 x\}$$

and target

$$T(X) = H^{n+3}(X; Z/2)/Sq^2(H^{n+1}(X)) + Sq^3(H^n(X)) \,.$$

Then our function

$$\Theta : S(X) \to T(X), \text{ on an element } x,$$

is given by the image in $T(X)$ of the element

$$(\tilde{\varphi} \circ \bar{\epsilon})^* b_{n+3}, \quad \text{where } \epsilon \text{ represents } x \,.$$

It will transpire that the formation of the quotient resolves any ambiguity in the choice of $\bar{\epsilon}$. The "secondary cohomology operation" Θ enjoys the same naturality properties as primary operations and makes sense in spite of $\theta, \varphi, \epsilon$ being defined only up to homotopy.

Having constructed an operation, we are faced with the problem of its calculation. The "secondary compositions" $\tilde{\varphi} \circ \bar{\epsilon}$ are key to our method. There is an adjoint relationship (with a sign) between the two sorts of secondary compositions,

$$\begin{gathered}(\text{extension}) \circ (\text{coextension})\,, \\ (\text{colifting}) \circ (\text{lifting})\end{gathered}$$

which has wide scope and produces results.

Often, the results we seek are encountered in a situation which is typical for many parts of homotopy theory, where calculations are made. Start with a diagram of the form

$$\begin{array}{ccccc} A & \xrightarrow{\alpha} & B & \xrightarrow{\beta} & C \\ & & {\scriptstyle \epsilon}\downarrow & & \downarrow{\scriptstyle \zeta} \\ & & X & \xrightarrow[f]{} & Y & \xrightarrow[g]{} & Z \end{array}$$

with $\beta\alpha \sim *$, $gf \sim *$, $\zeta\beta \simeq f\epsilon$. Then this diagram may be embedded in the following homotopy commutative diagram,

$$
\begin{array}{cccccccccccc}
 & & & & & & & & \Sigma A & = & \Sigma A \\
 & & & & & & & & \downarrow & & \downarrow{\scriptstyle \Sigma\alpha} \\
 & & A & \xrightarrow{\alpha} & B & \xrightarrow{\beta} & C & \longrightarrow & T_\beta & \longrightarrow & \Sigma B \\
 & & \downarrow & & \downarrow{\scriptstyle \epsilon} & & {\scriptstyle \zeta}\downarrow & & \downarrow & & \\
\Omega Y & \longrightarrow & W_f & \longrightarrow & X & \xrightarrow[f]{} & Y & \xrightarrow[g]{} & Z & & \\
{\scriptstyle \Omega g}\downarrow & & \downarrow & & & & & & & & \\
\Omega Z & = & \Omega Z\,, & & & & & & & &
\end{array}
$$

where the vertical compositions are adjoint up to sign. This construction is developed in Chapter 3. For the moment, we note that the vertical composition on the left represents the value of a secondary operation. The main strategy behind our calculations is to make inferences using the information represented on the right side of the diagram. We pass to the foundations.

Chapter 3

Fundamental Constructions

Much of the geometry for this subject is developed from a pair of maps and a null homotopy for their composition. The constructions are elementary, but require choices. We work in the category $\mathcal{K}_*$ of compactly generated spaces with non-degenerate basepoint, as in [**117**]. In particular, maps and homotopies between them are pointed; and cones, suspensions, etc. are reduced.

3.1. Terminology and conventions

3.1.1. Reduced cone. Let $(X, *)$ be a space with basepoint. Choose 0 for the basepoint of the unit interval I. The reduced cone TX is defined as $TX = I \wedge X$. As a quotient space

$$TX = I \times X/\{0\} \times X \cup I \times \{*\} .$$

We regard X in TX as the image of $\{1\} \times X$.

If $f : X \to Y$ is null-homotopic in $\mathcal{K}_*$, then a homotopy

$$H : I \times X \to Y \, , \; H(0, x) = * \, , \; H(1, x) = f(x) \, ,$$

induces a map (which we shall also write as H)

$$H : TX \to Y$$

which extends f. We say that H is a <u>contracting homotopy</u> from $*$ to f.

3.1.2. Adjoints of maps. We write the adjoint of

$$f : I \times X \to Y \text{ as } f^\natural : X \to Y^I,\ f^\natural(x)(t) = f(t,x);$$

and

$$g : X \to Y^I \text{ as } g^\flat : I \times X \to Y,\ g^\flat(t,x) = g(x)(t)\ .$$

The same notation will be used when suspensions and loop spaces are involved.

3.1.3. Space of paths. We write PX for the space of pointed maps from I to X. Thus

$$PX = \{w : I \to X \mid w(0) = *\}\ .$$

Given a contracting homotopy from I to $f : X \to Y$, we have

$$H^\natural : X \to PY,\ H^\natural(x)(s) = H(s,x)\ .$$

We also have a fibration

$$e : PX \to X\ ,\ e(w) = w(1)\ .$$

3.1.4. Fiber square. A fiber square is a pull-back diagram

$$\begin{array}{ccc} E & \xrightarrow{h} & E_0 \\ {\scriptstyle p}\downarrow & & \downarrow{\scriptstyle p_0} \\ B & \xrightarrow{f} & B_0\ , \end{array}$$

where p_0 is a fibration, as defined in [**117**]. If the loop functor is applied to a fiber square, the result is a fiber square

$$\begin{array}{ccc} \Omega E & \xrightarrow{\Omega h} & \Omega E_0 \\ {\scriptstyle \Omega p}\downarrow & & \downarrow{\scriptstyle \Omega p_0} \\ \Omega B & \xrightarrow[\Omega f]{} & \Omega B_0\ . \end{array}$$

We write $F = p_0^{-1}(*)$ for the fiber of p_0.

3.1.5. Principal action. If the space E_0 in (1.4) accepts a map

$$\mu_0 : F \times E_0 \to E_0$$

such that the diagram

$$\begin{array}{ccc} F \times E_0 & \xrightarrow{\mu_0} & E_0 \\ {\scriptstyle r_2}\downarrow & & \downarrow{\scriptstyle p_0} \\ E_0 & \xrightarrow[p_0]{} & B_0 \end{array}$$

commutes, where r_2 is projection to the second factor, then there is an induced action

$$\mu : F \times E \to E\,,\ \mu(x,y) = (p(y), \mu_0(x, h(y)))\ .$$

3.1.6. Homotopy fiber of a map. Given $f : B \to B_0$, we construct the fiber square

$$\begin{array}{ccc} W_f & \xrightarrow{h} & PB_0 \\ {\scriptstyle p}\downarrow & & \downarrow{\scriptstyle e} \\ B & \xrightarrow{f} & B_0 \end{array}$$

called the <u>homotopy fiber</u> of f. Elements of W_f are pairs

$$(b, w) \text{ in } B \times PB_0 \text{ such that } f(b) = w(1)\ .$$

It is a small point, but W_f differs from the "mapping fiber", T^f as defined in [**117**]. This space is constructed by "turning f into a fibration" and taking the fiber of the result. It is homeomorphic to our W_f, but the canonical homeomorphism reverses the direction of the loops in the fiber ΩB_0.

We have

$$\mu_0 : \Omega B_0 \times PB_0 \to PB_0$$

defined by

$$\mu_0(\sigma, \omega)(t) = \begin{cases} \sigma(2t) & 0 \le t \le \frac{1}{2}\,, \\ \omega(2t-1) & \frac{1}{2} \le t \le 1\,. \end{cases}$$

Then μ_0 defines a principal action of ΩB_0 on W_f which we write

$$\mu : \Omega B_0 \times W_f \to W_f\ .$$

Next we compare W_f with the pull-back from a contractible fibration. Suppose we have a fiber square

$$\begin{array}{ccc} W & \xrightarrow{h} & E_0 \\ {\scriptstyle p}\downarrow & & \downarrow{\scriptstyle p_0} \\ B & \xrightarrow[f]{} & B_0 \end{array}$$

with E_0 contractible to a point. More precisely, let L be a contracting homotopy for E_0

$$L : TE_0 \to E_0\ .$$

The adjoint of L composed with p_0 yields a map, such that the diagram

$$\begin{array}{ccccc} E_0 & \xrightarrow{L^\natural} & PE_0 & \xrightarrow{P(p_0)} & PB_0 \\ \Big\downarrow{\scriptstyle p_0} & & & & \Big\downarrow{\scriptstyle e} \\ B_0 & & \overline{\overline{\qquad\qquad}} & & B_0 \end{array}$$

commutes. The fibration p is fiber homotopy equivalent to the homotopy fiber of the map f, by naturality, in the following diagram:

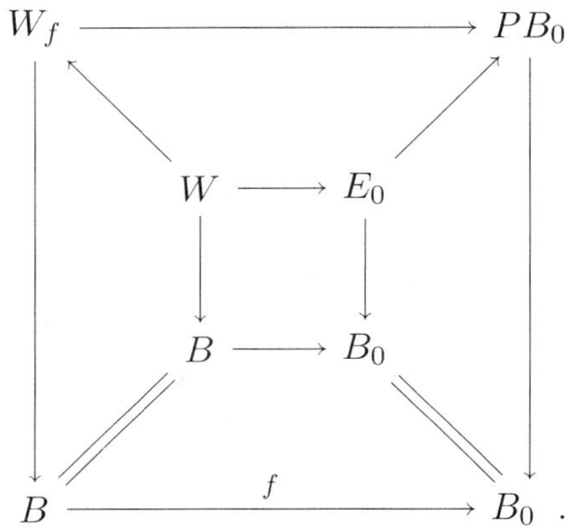

3.1.7. Loop space of W_f. Let $\kappa : \Omega PB_0 \to P\Omega B_0$ be the "switch of variables" map. We describe κ in terms of elements. An element of ΩPB_0 can be regarded as a map

$$\omega : I \times I \to B_0 \text{ such that}$$
$$\omega(t,0) = * \ , \ \omega(0,s) = * \text{ and } \omega(1,s) = * \ .$$

Then $\kappa(\omega)$ is the map

$$\kappa(\omega)(t,s) = \kappa(s,t)$$

with the variables switched. It is an exercise to check that κ induces multiplication by -1 on the homotopy groups of $\Omega^2 B_0$. Furthermore, κ induces a map of fiber squares

$$\begin{array}{ccc} \Omega W_f & \xrightarrow{\Omega h} & \Omega PB_0 \\ \Big\downarrow{\scriptstyle \Omega p} & & \Big\downarrow{\scriptstyle \Omega e} \\ \Omega B & \xrightarrow[\Omega f]{} & \Omega B_0 \end{array} \quad \text{to} \quad \begin{array}{ccc} W_{\Omega f} & \longrightarrow & P\Omega B_0 \\ \Big\downarrow{\scriptstyle q} & & \Big\downarrow \\ \Omega B & \xrightarrow[\Omega f]{} & \Omega B_0 \end{array}$$

which is a homeomorphism $\bar{\kappa}$ of the total spaces and

$$\begin{array}{ccccc} \Omega^2 B_0 & \longrightarrow & \Omega W_f & \xrightarrow{\Omega p} & \Omega B \\ \downarrow{\scriptstyle \kappa} & & \downarrow{\scriptstyle \bar{\kappa}} & & \| \\ \Omega^2 B_0 & \longrightarrow & W_{\Omega f} & \xrightarrow[q]{} & \Omega B \end{array} \qquad \text{commutes.}$$

The construction in (1.6) may be applied to produce $\bar{\kappa}$ from a contracting homotopy. Consider the homotopy

$$L : I \times PB_0 \to PB_0$$

given by $L(t, \omega)(s) = \omega(ts)$.

We define a homotopy

$$\Omega L : I \times \Omega PB_0 \to \Omega PB_0$$

by

$$\Omega L(t, \omega)(s_1, s_2) = \omega(s_1, ts_2) \ .$$

The composite of ΩL with Ωe is obtained by setting $s_2 = 1$. Then the switch of variables map is produced by taking the adjoint

$$\kappa = (\Omega e \circ \Omega L)^\natural$$

and $\bar{\kappa}$ is produced by naturality.

3.1.8. Mapping cone. Given $f : X \to Y$, we define the mapping cone T_f as the push-out in the diagram

$$\begin{array}{ccc} X & \longrightarrow & TX \\ \downarrow & & \downarrow \\ Y & \longrightarrow & T_f \ . \end{array}$$

This construction agrees with [**117**] so we use the same notation.

This concludes our recollection of terminology. We assume the reader is familiar with the standard material concerning these notions, as found in [**117**].

3.2. Basic constructions

3.2.1. Maps constructed from contracting homotopies. The constructions discussed in this section are managed by means of naturality arguments for fiber squares and push-outs. We first describe a space X in these terms. We regard X as a pull-back using either the diagram

$$\begin{array}{ccc} X & = & X \\ \downarrow & & \downarrow \\ * & = & * \end{array}$$

or, identifying X with the diagonal Δ in $X \times X$,

$$\begin{array}{ccc} \Delta & \xrightarrow{r_2} & X \\ {\scriptstyle r_1}\downarrow & & \| \\ X & = & X \end{array}$$

where r_1, r_2 are the projections to the first and second factors respectively. To display X as a push-out, we can use either of the following diagrams:

$$\begin{array}{ccc} * & \longrightarrow & X \\ \downarrow & & \| \\ * & \longrightarrow & X \end{array} \qquad \text{or} \qquad \begin{array}{ccc} X & \longrightarrow & X \\ \| & & \| \\ X & \longrightarrow & X \end{array} .$$

The <u>reduced suspension</u> $\Sigma X = S^1 \wedge X$ where S^1 is obtained from I by identifying $\{0\}$ and $\{1\}$. We express the reduced suspension as a push-out, up to change of suspension coordinate in the following manner. Let T_0X be the cone from (1.1) and T_1X be $I \wedge X$ where $\{1\}$ is the base point for I. Then we have the diagram

$$\begin{array}{ccc} X & \longrightarrow & T_0X \\ \downarrow & & \downarrow \\ T_1X & \longrightarrow & \Sigma X \end{array} \qquad \text{with maps:} \qquad \begin{array}{ccc} x & \rightarrowtail & (1, x), \\ \downarrow & & \\ (0, x) & & \end{array} \qquad \begin{array}{c} (t, x) \\ \downarrow \\ (t/2, x) \end{array}$$

$$(t, x) \rightarrowtail \left(\frac{t+1}{2}, x\right)$$

for $0 \leq t \leq 1$.

Similarly, let P_0X be the space of paths from (1.3) and let P_1X be the space of paths where $\{1\}$ is the basepoint for I. Then we can produce the loop space ΩX, in a pull-back diagram, up to change of the path coordinate,

$$\begin{array}{ccc} \Omega X & \xrightarrow{i_0} & P_0 X \\ {\scriptstyle i_1}\downarrow & & \downarrow{\scriptstyle e_0} \\ P_1 X & \xrightarrow[e_1]{} & X \end{array} \qquad \text{with maps} \qquad \begin{aligned} (i_0\omega)(t) &= \omega(t/2) \ , \\ (i_1\omega)(t) &= \omega\left(\frac{t+1}{2}\right) \ , \\ e_0(\omega) &= \omega(1) \ , \\ e_1(\omega) &= \omega(0) \ . \end{aligned}$$

The <u>direction reversal</u> map of the unit interval is given by

$$\tau : I \to I \ , \ \tau(t) = 1 - t \ .$$

When τ is used to reverse direction in homotopies, we write

$$H_\tau(t, x) = H(\tau(t), x) \ .$$

Similarly, given $f : X \to \Omega Y$,

$$f_\tau : X \to \Omega Y \text{ is given by } f_\tau(x)(s) = f(x)(\tau(s)) \ ;$$

and for $f : \Sigma X \to Y$,

$$f_\tau : \Sigma X \to Y \text{ is given by } f_\tau(t, x) = f(\tau(t), x) \ .$$

A <u>sequence with homotopy</u> (β, α, H) is a pair of composable maps

$$A \xrightarrow{\alpha} B \xrightarrow{\beta} C$$

and a contracting homotopy H from $*$ to $\beta\alpha$. Starting with this data, we have four new maps.

3.2.2. Extension of $\underline{\beta} : T_\alpha \to C$. This map is obtained by naturality from the push-out data,

$$\underline{\beta} : \quad \begin{array}{ccccc} B & \xleftarrow{\alpha} & A & \longrightarrow & T_0 A \\ {\scriptstyle \beta}\downarrow & & \downarrow{\scriptstyle \beta\alpha} & & \downarrow{\scriptstyle H} \\ C & = & C & = & C \ . \end{array}$$

3.2.3. Coextension of $\underset{\sim}{\alpha} : \Sigma A \to T_\beta$. This map is obtained by naturality from the push-out data,

$$\underset{\sim}{\alpha} : \quad \begin{array}{ccccc} T_1 A & \longleftarrow & A & \longrightarrow & T_0 A \\ {\scriptstyle H_\tau}\downarrow & & \downarrow{\scriptstyle \alpha} & & \downarrow{\scriptstyle T_0\alpha} \\ C & \xleftarrow[\beta]{} & B & \longrightarrow & T_0 B \end{array}$$

and note the direction reversal map.

3.2.4. Lifting of $\alpha : A \to W_\beta$. This map is obtained by naturality from pull-back data,

$$\bar{\alpha} : \quad \begin{array}{ccccc} A & = & A & = & A \\ {\scriptstyle \alpha}\downarrow & & \downarrow{\scriptstyle \beta\alpha} & & \downarrow{\scriptstyle H^\natural} \\ B & \xrightarrow[\beta]{} & C & \longleftarrow & P_0 C \end{array}$$

and note the adjoint of H.

3.2.5. Colifting of $\beta : W_\alpha \to \Omega C$. This map is obtained by naturality from pull-back data,

$$\tilde{\beta} : \quad \begin{array}{ccccc} A & \xrightarrow{\alpha} & B & \longleftarrow & P_0 B \\ {\scriptstyle H_\tau^\natural}\downarrow & & \downarrow{\scriptstyle \beta} & & \downarrow{\scriptstyle P_0\beta} \\ P_1 C & \longrightarrow & C & \longleftarrow & P_0 C \end{array}$$

and note both the adjoint and direction reversal.

Given three composable maps

$$A \xrightarrow{\alpha} B \xrightarrow{\beta} C \xrightarrow{\gamma} D$$

and contracting homotopies H from $*$ to $\beta\alpha$, K from $*$ to $\gamma\beta$, we can construct two *secondary compositions*,

$$\underline{\gamma} \circ \underset{\sim}{\alpha} : \Sigma A \to T_\beta \to D \quad \text{through the mapping cone}$$

and

$$\tilde{\gamma} \circ \bar{\alpha} : A \to W_\beta \to \Omega D \quad \text{through the homotopy fiber.}$$

Lemma 3.2.1 (Interface Lemma). There is equality of maps

$$(\underline{\gamma} \circ \underset{\sim}{\alpha})^\natural = (\tilde{\gamma} \circ \bar{\alpha})_\tau\,, \quad \text{note the direction reversal.}$$

Proof. The composition $\underline{\gamma} \circ \underset{\sim}{\alpha}$ is obtained by fitting the diagrams in (3.2.2), (3.2.3) together,

$$\begin{array}{ccccc} T_1 A & \longleftarrow & A & \longrightarrow & T_0 A \\ {\scriptstyle H_\tau}\downarrow & & \downarrow{\scriptstyle \alpha} & & \downarrow{\scriptstyle T_0\alpha} \\ C & \xleftarrow{\beta} & B & \longrightarrow & T_0 B \\ {\scriptstyle \gamma}\downarrow & & \downarrow{\scriptstyle \gamma\beta} & & \downarrow{\scriptstyle K} \\ D & = & D & = & D\,. \end{array}$$

Likewise, the composition $\tilde{\gamma}\circ\bar{\alpha}$ is obtained by fitting the diagrams in (3.2.4), (3.2.5) together,

$$\begin{array}{ccccc}
A & = & A & = & A \\
{\scriptstyle\alpha}\downarrow & & \downarrow{\scriptstyle\beta\alpha} & & \downarrow{\scriptstyle H^\natural} \\
B & \xrightarrow[\beta]{} & C & \longleftarrow & P_0C \\
{\scriptstyle K_\tau^\natural}\downarrow & & \downarrow{\scriptstyle\gamma} & & \downarrow{\scriptstyle P_0\gamma} \\
P_1D & \longrightarrow & D & \longleftarrow & P_0D
\end{array} .$$

The result follows by inspection. □

The lemma is so named to convey its mediating role between constructions in the "world of cofibrations" and the "world of fibrations".

Before stating our next lemma, we introduce some notation used in its proof and elsewhere. Given a list of homotopies

$$H_i : I\times X \to Y\ ,\ 1\le i\le n$$

with $H_i(x,1)=H_{i+1}(x,0)$, we write

$$\{H_1,\dots,H_n\} : I\times X\to Y$$

given by

$$\{H_1,\dots,H_n\}(t,x)=H_i(x,nt-i+1) \text{ if } i-1\le nt\le i$$

to denote the homotopy obtained by joining the homotopies in the list.

The next lemma supplies the "move" needed for exploitation of the interface lemma.

Lemma 3.2.2 (Slide Lemma). In a diagram

$$\begin{array}{ccccccc}
A & \xrightarrow{\alpha} & B & \xrightarrow{\beta} & C & \xrightarrow{\varphi\zeta} & Z \\
\| & & \downarrow{\scriptstyle\epsilon} & & \downarrow{\scriptstyle\zeta} & & \| \\
A & \xrightarrow[\epsilon\alpha]{} & X & \xrightarrow[\theta]{} & Y & \xrightarrow[\varphi]{} & Z
\end{array}$$

with contracting homotopies H from $*$ to $\beta\alpha$ and K from $*$ to $\varphi\theta$, and a homotopy J from $\zeta\beta$ to $\theta\epsilon$, we have

$$\underline{\varphi\zeta}\circ\underset{\sim}{\alpha}\simeq\underline{\varphi}\circ\underset{\sim}{\epsilon\alpha}$$

for the extension/coextension composition and

$$\widetilde{\varphi\zeta}\circ\bar{\alpha}\simeq\tilde{\varphi}\circ\overline{\epsilon\alpha}$$

for the colifting/lifting composition.

Proof. The composition $\underline{\varphi\zeta} \circ \underset{\sim}{\alpha}$ is displayed,

$$
\begin{array}{ccccc}
T_1A & \longleftarrow & A & \longrightarrow & T_0A \\
{\scriptstyle H_\tau}\downarrow & & \downarrow{\scriptstyle \alpha} & & \downarrow{\scriptstyle T_0\alpha} \\
C & \overset{\beta}{\longleftarrow} & B & \longrightarrow & T_0B \\
{\scriptstyle \varphi\zeta}\downarrow & & \downarrow & & \downarrow{\scriptstyle \{K\epsilon,\varphi J_\tau\}} \\
Z & = & Z & = & Z
\end{array}
$$

and for the composition $\underline{\varphi} \circ \underline{\epsilon\alpha}$, we have

$$
\begin{array}{ccccc}
T_1A & \longleftarrow & A & \longrightarrow & T_0A \\
{\scriptstyle \{J_\tau\alpha,\zeta H_\tau\}}\downarrow & & \downarrow{\scriptstyle \epsilon\alpha} & & \downarrow{\scriptstyle T_0\epsilon\alpha} \\
Y & \overset{\theta}{\longleftarrow} & X & \longrightarrow & T_0X \\
{\scriptstyle \varphi}\downarrow & & \downarrow & & \downarrow{\scriptstyle K} \\
Z & = & Z & = & Z\ .
\end{array}
$$

The statement follows because

$$\{\{K\epsilon\alpha, \varphi J_\tau\alpha\}, \varphi\zeta H_\tau\} \simeq \{K\epsilon\alpha, \{\varphi J_\tau\alpha, \varphi\zeta H_\tau\}\} \ ,$$

by a homotopy in the suspension coordinate, and fixed on the space coordinate.

The proof for the other secondary composition is similar, or is a consequence of Lemma 3.2.1 and the first part of this lemma. □

Exercise 3.2.3. Verify the following formulas for the constructions in subheadings (3.2.2)–(3.2.5):

$$\underline{\beta}(t,a) = H(t,a)\,,\ \underline{\beta}(b) = \beta(b) \quad (t,a) \text{ in } T_0A, b \text{ in } B\ ,$$

$$\underset{\sim}{\alpha}(t,a) = \begin{Bmatrix} (2t, \alpha(a)) & 0 \le t \le \frac{1}{2} \\ jH(2-2t, a) & \frac{1}{2} \le t \le 1 \end{Bmatrix}$$

where j is the inclusion of C in the mapping cone T_β.

$$\bar{\alpha}(a)(s) = (\alpha(a), H(s,a)) \ ,$$

$$\tilde{\beta}(a,\omega)(s) = \begin{Bmatrix} \beta\omega(2s) & 0 \le s \le \frac{1}{2} \\ H(2-2s,a) & \frac{1}{2} \le s \le 1 \end{Bmatrix} \ ;$$

$$\underline{\gamma} \circ \underset{\sim}{\alpha}(t,a) = \{K\alpha, \gamma H_\tau\} = \begin{Bmatrix} K(2t,\alpha(a)) & 0 \le t \le \frac{1}{2} \\ \gamma H(2-2t,a) & \frac{1}{2} \le t \le 1 \end{Bmatrix} \ ;$$

$$(\tilde{\gamma} \circ \bar{\alpha})(a)(s) = \{\gamma H, K_\tau \alpha\} = \begin{Bmatrix} \gamma H(2s,a) & 0 \le s \le \frac{1}{2} \\ K(2-2s,\alpha(a)) & \frac{1}{2} \le s \le 1 \end{Bmatrix} \ .$$

3.2.6. Remark. The fiber homotopy equivalence of (3.1.6) preserves co-iftings, in the following sense. Suppose we have a commutative diagram

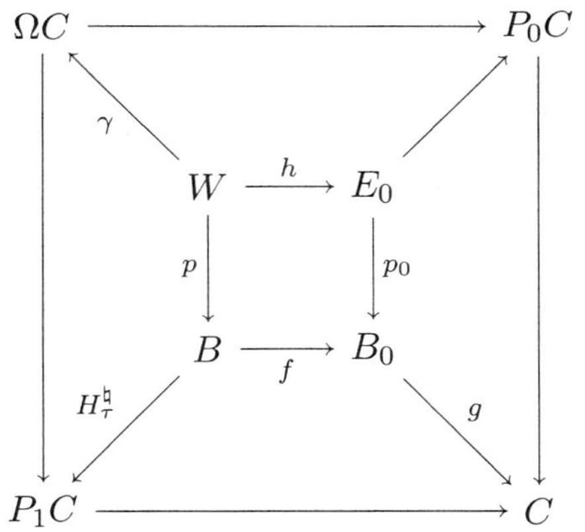

where the squares are fiber squares, E_0 is a contractible space and

$$(g, f, H)$$

is a sequence with homotopy. In particular, the map γ is induced by naturality. Moreover, the following diagram commutes,

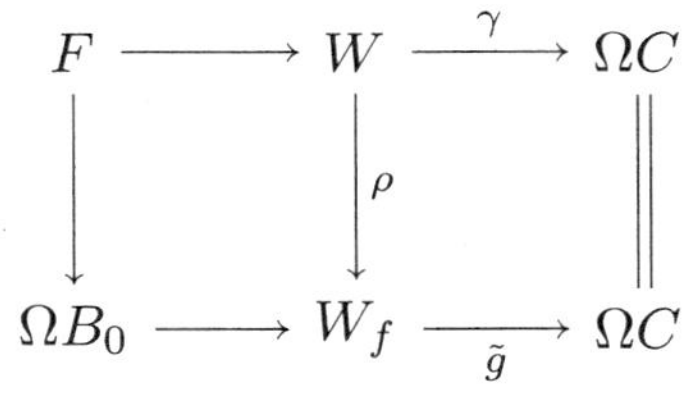

where F is the fiber of p_0 and ρ is a fiber homotopy equivalence constructed as in (3.1.6). This assertion follows from the commutative diagram,

$$\begin{array}{ccccc}
B & \xrightarrow{f} & B_0 & \xleftarrow{p_0} & E_0 \\
\| & & \| & & \downarrow{\scriptstyle P_0(p_0)\circ L^\natural} \\
B & \xrightarrow{f} & B_0 & \longleftarrow & P_0B_0 \\
{\scriptstyle H^\natural_\tau}\downarrow & & \downarrow{\scriptstyle g} & & \downarrow{\scriptstyle P_0(g)} \\
P_1C & \longrightarrow & C & \longleftarrow & P_0C\ .
\end{array}$$

3.2.7. Signs. In this section we address all questions of signs which arise in our work. Let

$$(\beta, \alpha, H)$$

be a sequence with homotopy for $\alpha : A \to B$, $\beta : B \to C$ and write

$$j : \Omega B \to W_\alpha \text{ given by } j(\omega) = (*, \omega)$$

for the inclusion of the fiber. The following diagrams commute up to homotopy,

$$\begin{array}{ccc}
\Omega B & \xrightarrow{\Omega\beta} & \Omega C \\
{\scriptstyle j}\downarrow & & \| \\
W_\alpha & \xrightarrow{\tilde\beta} & \Omega C \\
{\scriptstyle p}\downarrow & \boxed{-1} & \downarrow{\scriptstyle j} \\
A & \xrightarrow[\bar\alpha]{} & W_\beta \\
\| & & \downarrow \\
A & \xrightarrow[\alpha]{} & B
\end{array}
\quad , \quad
\begin{array}{ccc}
W_p & \xrightarrow{\tilde\alpha} & \Omega B \\
{\scriptstyle \hat p}\downarrow & \boxed{-1} & \downarrow{\scriptstyle j} \\
W_\alpha & = & W_\alpha
\end{array}$$

except in the middle, where

$$\bar\alpha \circ p \simeq j \circ \tilde\beta_\tau\ ,\ \hat p \simeq j \circ \tilde\alpha_\tau\ .$$

The square on the right is the special case for (α, p, H), with

$$H(s, (a, \omega)) = \omega(s)\ .$$

Likewise, writing

$$\begin{aligned}
j :& B \to T_\alpha\ ,\ \text{inclusion at the base of the cone}\ ,\\
q :& T_\alpha \to \Sigma A\ ,\ \text{collapse } B \text{ to a point;}
\end{aligned}$$

then the following diagrams commute up to homotopy,

$$
\begin{array}{ccc}
B & \xrightarrow{\ \beta\ } & C \\
\downarrow{\scriptstyle j} & & \| \\
T_\alpha & \xrightarrow[\boxed{-1}]{\ \beta\ } & C \\
\downarrow{\scriptstyle q} & & \downarrow{\scriptstyle j} \\
\Sigma A & \xrightarrow[\ \underset{\sim}{\alpha}\ ]{} & T_\beta \ , \\
\| & & \downarrow \\
\Sigma A & \xrightarrow[\ \Sigma\alpha\ ]{} & \Sigma B
\end{array}
\qquad
\begin{array}{ccc}
T_\alpha & \xlongequal[\boxed{-1}]{} & T_\alpha \\
\downarrow{\scriptstyle q} & & \downarrow{\scriptstyle \hat{j}} \\
\Sigma A & \xrightarrow[\ \underset{\sim}{\alpha}\ ]{} & T_j
\end{array}
$$

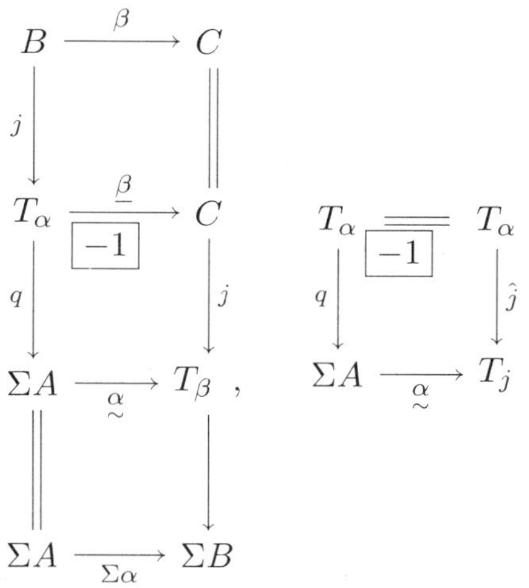

except in the middle, where

$$
j \circ \underline{\beta} \simeq \underset{\sim}{\alpha_\tau} \circ q \ , \ \hat{j} \simeq \underset{\sim}{\alpha_\tau} \circ q \ .
$$

The square on the right is the special case for (j, α, H), with $H(t, a) = (t, a)$.

A homotopy

$$
L : I \times W_\alpha \to W_\beta
$$

from $j \circ \tilde{\beta}_\tau$ to $\bar{\alpha} \circ p$ is given by

$$
L(t, a, \omega) = \left(\omega(t), \left\{ \begin{array}{l} H\left(\dfrac{2s}{t+1} \ , \ a \right) 0 \le t \le 1 \ , \ 0 \le s \le \dfrac{t+1}{2} \\[2ex] \beta\omega(2 + t - 2s) 0 \le t \le 1 \ , \ \dfrac{t+1}{2} \le s \le 1 \end{array} \right\} \right) .
$$

We leave the other details to the reader.

The signs discussed above are essentially those appearing in [**117**], Chapter 3. Some people call it "the sign that won't go away". We call it the *compatibility* sign when it is encountered in discussions of our fundamental constructions. For iterations of the mapping cone or homotopy fiber constructions, we use the above relations to make an *ad hoc* determination of sign.

Remark. For the case where (B, A) is an NDR pair and α is the inclusion of the subspace A into B, we compare the map $r : B/A \to \Sigma A$ from subsection 1.3.3 with the map q. Recall that r is the homotopy inverse to a map collapsing $T_1 A$ to a point in $B \cup T_1 A$. If we collapse $TA = T_0 A$ to a point in T_α, we obtain a homotopy equivalence $k : T_\alpha \to B/A$ and kj is the projection from B to B/A. We can compare r with q through the direction reversal map on ΣA. The result is

$$rk \simeq (Id)_\tau \circ q\ .$$

We next consider signs that arise when the loop functor Ω is applied to the lifting and colifting constructions. We combine (3.1.7) with the lifting and colifting constructions. For a sequence with homotopy, (β, α, H), we have another $(\Omega\beta, \Omega\alpha, \Omega H)$ where

$$\begin{gathered}\Omega H : I \times \Omega A \to \Omega C \text{ is given by } \Omega H(t,\omega) = H(t,\omega)\\ \text{for } \omega \text{ in } \Omega A, \text{ thus } \Omega H(t,\omega)(s) = H(t,\omega(s))\ .\end{gathered}$$

Now, we have the equation for adjoints,

$$(\Omega H)^\natural = \kappa \circ \Omega(H^\natural)\ ,\ \kappa = \text{ switch of variables.}$$

This equation is evident from the formulas obtained by fixing ω in ΩA, and regarding the results as a pair of maps

$$I \times I \to C\ .$$

The formulas are,

$$(\Omega H)^\natural(\omega)(x,y) = H(x,\omega(y))$$

and

$$\Omega(H^\natural)(\omega(x,y) = H(y,\omega(x))\ .$$

Thus, when Ω is applied to a lifting, we have

$$\bar\kappa \circ \Omega\bar\alpha = \overline{\Omega\alpha}$$

and for a colifting,

$$\kappa \circ \Omega\tilde\beta = \widetilde{\Omega\beta} \circ \bar\kappa \quad (\bar\kappa \text{ as in (3.1.7)})$$

for liftings and coliftings constructed from (β, α, H), $(\Omega\beta, \Omega\alpha, \Omega H)$.

For some discussions, the appearance of the switch of variables is a nuisance. We deal with it by a construction called *replacement of a loop of a colifting by a colifting compatible with the inclusion of the fiber.*

We establish that there is a fiber homotopy equivalence between the fibrations Ωp and q, from the fiber squares,

$$\begin{array}{ccc} \Omega W_\alpha & \longrightarrow & \Omega PB \\ {\scriptstyle \Omega p}\downarrow & & \downarrow \\ \Omega A & \xrightarrow{\Omega\alpha} & \Omega B \end{array} \qquad \text{and} \qquad \begin{array}{ccc} W_{\tau\circ\Omega\alpha} & \longrightarrow & P\Omega B \\ {\scriptstyle q}\downarrow & & \downarrow \\ \Omega A & \xrightarrow{\tau\circ\Omega\alpha} & \Omega B \end{array}\ ,$$

where τ is the direction reversal map. Furthermore, the following diagram commutes up to homotopy,

$$\begin{array}{ccccc} \Omega^2 B & \xrightarrow{\Omega j} & \Omega W_\alpha & \xrightarrow{\Omega\tilde{\beta}} & \Omega^2 C \\ \Big\| & & \Big\downarrow{\scriptstyle \rho} & & \Big\downarrow{\scriptstyle \tau\circ\kappa\simeq Id.} \\ \Omega^2 B & \xrightarrow[j]{} & W_{\tau\circ\Omega\alpha} & \xrightarrow[\widetilde{\Omega\beta}]{} & \Omega^2 C\ , \end{array}$$

where the colifting $\widetilde{\Omega\beta}$ is constructed from $(\Omega\beta, \tau\circ\Omega\alpha, \tau\circ\Omega H)$ and the fiber homotopy equivalence ρ is constructed using τ and κ, as in the commutative diagram below:

$$\begin{array}{ccccccccc} \Omega A & & \xrightarrow{\Omega\alpha} & & \Omega B & & \longleftarrow & & \Omega P_0 B \\ \Big\downarrow & \overset{=}{\searrow} & & & \Big\downarrow{\scriptstyle \tau} & & & \underset{\tau\circ\kappa}{\swarrow} & \Big\downarrow \\ & & \Omega A & \xrightarrow{\tau\circ\Omega\alpha} & \Omega B & \longleftarrow & P_0\Omega B & & \\ {\scriptstyle \Omega(H_\tau^\natural)}\Big\downarrow & & {\scriptstyle (\tau\circ\Omega H_\tau)^\natural}\Big\downarrow & & \Big\downarrow{\scriptstyle \Omega\beta} & & \Big\downarrow{\scriptstyle P_0\Omega\beta} & & \Big\downarrow{\scriptstyle \Omega P_0\beta} \\ & & P_1\Omega C & \longrightarrow & \Omega C & \longleftarrow & P_0\Omega C & & \\ \Big\downarrow & \underset{\tau\circ\kappa}{\nearrow} & & & \Big\uparrow{\scriptstyle \tau} & & & \underset{\tau\circ\kappa}{\nwarrow} & \Big\downarrow \\ \Omega P_1 C & & \longrightarrow & & \Omega C & & \longleftarrow & & \Omega P_0 C\ . \end{array}$$

We observe in general that the restriction to $\Omega^2 X$ of the map

$$\tau\circ\kappa : \Omega P X \to P\Omega X$$

with τ acting on the loop variable, is canonically homotopic to the identity. Regarded as maps from $I\times I$, elements of $\Omega^2 X$ are mapped by $\tau\circ\kappa$ by the formula

$$\tau\circ\kappa(\omega)(x,y) = \omega(y, 1-x)\ .$$

This symmetry of the square $I\times I$ is clockwise rotation by 90 degrees. Hence a common homotopy can be used for all spaces X.

Remark. We write $\rho = \rho_\alpha$ to indicate the dependence on α, if more than one map is involved. The fiber homotopy equivalence ρ is determined by τ and κ, thus it enjoys some obvious naturality properties.

3.2.8. Compatibility of coliftings with the principal action. Here we compose a colifting with the principal action. The following diagram commutes up to homotopy,

$$\begin{array}{ccc} \Omega B \times W_\alpha & \xrightarrow{\mu} & W_\alpha \\ \Big\downarrow{\scriptstyle \Omega\beta\times\tilde{\beta}} & & \Big\downarrow{\scriptstyle \tilde{\beta}} \\ \Omega C \times \Omega C & \xrightarrow{+} & \Omega C \end{array}$$

where $\tilde{\beta}$ is constructed from (β, α, H) and $+$ denotes loop addition. This fact is seen by combining the formula in (3.1.5) with one from exercise (3.2.3). The significance of (3.2.8) is that maps from W_α represented by coliftings behave especially well under the principal action.

3.3. Dependence on homotopies

Our constructions are given in terms of maps and homotopies. First we discuss the dependence of the results on the choice of homotopy. Next, we establish invariance properties for secondary compositions.

The reader who is familiar with the "toral" or Toda "bracket" construction may wonder why we are not passing immediately to that construction. Our answer is that this step casts aside important information. More remarks along this line appear in section 3.3.3.

3.3.1. Alteration of contracting homotopies. Let H_1, H_2 be contracting homotopies for the composition $\beta\alpha$. We measure their difference by

$$\delta(H_1, H_2) : A \to \Omega C \text{ given by}$$

$$\delta(H_1, H_2) = \{H_1, H_{2\tau}\}^\natural = \begin{Bmatrix} H_1(2s, a) & 0 \le s \le \frac{1}{2} \\ H_2(2-2s, a) & \frac{1}{2} \le s \le 1 \end{Bmatrix}.$$

Let

$$\bar{\alpha}_i : A \to W_\beta \quad i = 1, 2$$

be the corresponding liftings. Then we obtain

$$\bar{\alpha}_1 \simeq \mu \circ (\delta(H_1, H_2) \times \bar{\alpha}_2) \circ \Delta$$

where Δ is the diagonal map on A. This can be seen by combining the formula in (3.1.5) with a formula from exercise (3.2.3).

Let $\tilde{\beta}_i : W_\alpha \to \Omega C \quad i = 1, 2$ be the corresponding coliftings. Then

$$\tilde{\beta}_2 \simeq \tilde{\beta}_1 + \delta(H_1, H_2) \circ p$$

where $p : W_\alpha \to A$ is the projection and the sum is loop addition. This relation is evident from (3.2.3).

Given maps

$$A \xrightarrow{\alpha} B \xrightarrow{\beta} C \xrightarrow{\gamma} D$$

with a pair H_1, H_2 of contracting homotopies for $\beta\alpha$ and a contracting homotopy κ for $\gamma\beta$, we have

$$\tilde{\gamma} \circ \bar{\alpha}_1 \simeq \Omega\gamma \circ \delta(H_1, H_2) + \tilde{\gamma} \circ \bar{\alpha}_2$$

using subsection (3.2.8). If instead, we have a contracting homotopy H for $\beta\alpha$ and a pair K_1, K_2 of contracting homotopies for $\gamma\beta$, then we have

$$\tilde{\gamma}_2 \circ \bar{\alpha} \simeq \tilde{\gamma}_1 \circ \bar{\alpha} + \delta(K_1, K_2) \circ \alpha \ .$$

3.3.2. Strong invariance of coliftings. We observe that $\tilde{\beta}_1$ is homotopic to $\tilde{\beta}_2$ if $\delta(H_1, H_2)$ is null-homotopic. There is a stronger form of invariance. To describe it, we introduce the *Half-smash product* construction.

Definition 3.3.1. Let A be a space and B be a pointed space. Then the *half-smash product* is given by

$$A \ltimes B = A^+ \wedge B$$

where A^+ is A with a disjoint basepoint.

As a quotient

$$A \ltimes B = A \times B / A \times \{*\}$$

where $*$ is the basepoint in B.

If A is already pointed, then we have canonical maps

$$S^0 \to A^+ \ , \ 0 \mapsto * \text{ of } A \ , \ 1 \mapsto +$$

and

$$A^+ \to S^0 \ , \{A\} \to 0, + \to 1 \ .$$

Thus B is a retract of $A \ltimes B$, via the composition

$$B = S^0 \wedge B \to A^+ \wedge B \to S^0 \wedge B \ .$$

Furthermore, the identity map of A extends to A^+ by sending the disjoint basepoint $+$ to the basepoint $*$ in A. Thus we have a projection

$$A \ltimes B \to A \wedge B \ .$$

The retraction map yields a split monomorphism in the homology sequence for the pair $(A \ltimes B, B)$, thus

$$\bar{H}_*(A \ltimes B) \cong \bar{H}_*(B) \oplus \bar{H}_*(A \wedge B) \ .$$

More information on this construction is given as exercises.

Exercise 3.3.2. Use cup products to distinguish the homotopy types of

$$CP^2 \ltimes CP^2 \text{ and } CP^2 \vee CP^2 \wedge CP^2 \ .$$

Exercise 3.3.3. If A and B are connected CW complexes, prove

$$\Sigma(A \ltimes B) \simeq \Sigma B \vee \Sigma(A \wedge B)\ .$$

Exercise 3.3.4. If B is a co-H-space or a cogroup, prove that the same is true for $A \ltimes B$.

The strong invariance property is

Proposition 3.3.5. Let H_1, H_2 be contracting homotopies for the composition $\beta\alpha$. If $\delta(H_1, H_2)$ factors through $A \ltimes B$ up to homotopy, then $\tilde{\beta}_1$ is homotopic to $\tilde{\beta}_2$,

$$\begin{array}{ccc} A & \xrightarrow{\delta} & \Omega C \\ {\scriptstyle 1\ltimes\alpha}\downarrow & \nearrow_{\delta'} & \\ A \ltimes B & & \end{array}\ , \qquad 1 \ltimes \alpha(a) = \text{ image of } (a, \alpha(a)) \text{ in } A \ltimes B\ .$$

Proof. We have the commutative diagram

$$\begin{array}{ccccc} W_\alpha & \longrightarrow & W_\alpha \times PB & \longrightarrow & W_\alpha \ltimes PB \\ {\scriptstyle p}\downarrow & & \downarrow{\scriptstyle p\times e} & & \downarrow{\scriptstyle p\ltimes e} \\ A & \xrightarrow[1\times\alpha]{} & A \times B & \longrightarrow & A \ltimes B \end{array}$$

Hence $\delta \circ p$ factors through $W_\alpha \ltimes PB$, which is contractible. □

3.3.3. Homotopy invariance for secondary compositions. Starting with a pair of sequences with homotopy

$$(\beta, \alpha, H)\ ,\ (\gamma, \beta, K)$$

we can project $\tilde{\gamma} \circ \bar{\alpha}$ into the double quotient where we assume, for convenience, that D is a simply-connected H-space,

$$im\alpha^{\#} : [B, \Omega D] \to [A, \Omega D] \setminus [A, \Omega D]/im\Omega\gamma_{\#} : [A, \Omega C] \to [A, \Omega D]\ .$$

The formulas in (3.3.1) show that the result is independent of the contracting homotopies. It follows easily from this, that the result is an invariant of the homotopy classes of the maps α, β, γ. Now dividing by the image of $\Omega\gamma_{\#}$ will be acceptable; but division by the image of $\alpha^{\#}$ results in a less powerful theory for secondary cohomology operations. An illustration is found in the remark after Prop. 5.1.4 in Chapter 5.

In this section, we analyze the influence on a secondary composition by variation within the homotopy class of each factor. Let α, α' be maps $A \to B$, and β, β' be maps $B \to C$.

Definition 3.3.6. Two sequences with homotopy, (β, α, H) and (β', α', H'), are homotopic

$$(\beta, \alpha, H) \sim (\beta', \alpha', H')$$

provided there are homotopies L, M, from α to α', respectively β to β', such that

$$\delta(H', \{H, \beta L, M\alpha'\}) : A \to \Omega C$$

is null-homotopic.

The difference δ used in (3.3.1) is displayed schematically,

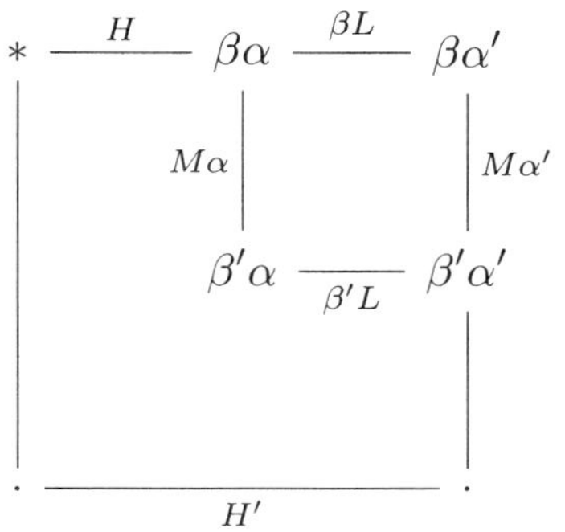

and is obtained by anti-clockwise traversal around the outside paths, starting and returning to the constant map $*$. The homotopy

$$ML : I \times I \times A \to C \text{ given by } ML(x, y, a) = M(x, L(y, a))$$

can be used to "fill in" the interior square. Hence

$$\delta(H', \{H, \beta L, M\alpha'\}) \simeq \delta(H', \{H, M\alpha, \beta' L\})$$

and either can be used in (3.3.6).

Proposition 3.3.7. Given pairs of maps $\alpha, \alpha' : A \to B; \beta, \beta' : B \to C$, and $\gamma, \gamma' : C \to D$ and contracting homotopies to form four sequences with homotopy

$$(\beta, \alpha, H), (\gamma, \beta, K),\ (\beta', \alpha', H'),\ (\gamma', \beta', K')$$

and given homotopies L from α to α', M from β to β' and N from γ to γ' such that

$$\delta_1 = \delta(H', \{H, M\alpha, \beta' L\}) : A \to \Omega C$$

and

$$\delta_2 = \delta(K', \{K, N\beta, \gamma' M\}) : B \to \Omega D$$

are null-homotopic, then the secondary compositions are homotopic,

$$\tilde{\gamma} \circ \bar{\alpha} \simeq \tilde{\gamma}' \circ \bar{\alpha}' \ ,$$

$$\underline{\gamma} \circ \underset{\sim}{\alpha} \simeq \underline{\gamma}' \circ \underset{\sim}{\alpha}' \ .$$

Proof. We write secondary compositions in terms of the difference construction, using exercise (3.2.3). Thus

$$\underline{\gamma} \circ \underset{\sim}{\alpha} = \delta(K\alpha, \gamma H) ,$$
$$\tilde{\gamma} \circ \bar{\alpha} = \delta(\gamma H, K\alpha) .$$

Then, the required homotopies are evident by inspection of the following schematic diagram:

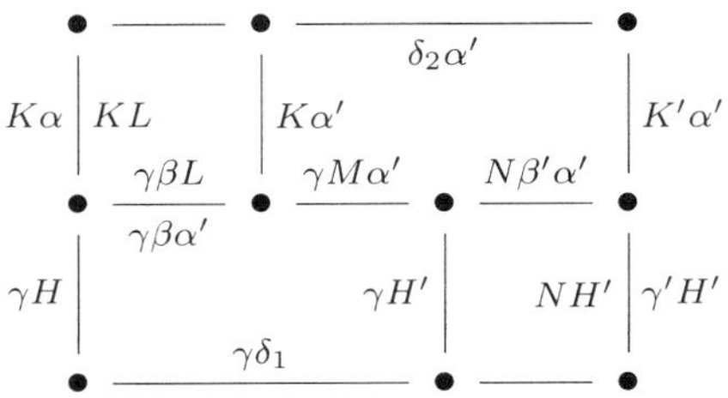

□

Remark. Proposition 3.3.7 is often applied to $H' = \{H, M\alpha, \beta' L\}$ and $K' = \{K, N\beta, \gamma' M\}$, to produce a well-defined homotopy class $[\tilde{\gamma} \circ \bar{\alpha}]$ starting with homotopy classes $[\gamma], [\beta], [\alpha]$ and contracting homotopies for (composites of) representing maps.

3.4. Peterson-Stein formulas

3.4.1. A key pattern. A structure found in many calculations is the following array of spaces and maps,

$$\begin{array}{ccccc} A & \xrightarrow{\alpha} & B & \xrightarrow{\beta} & C & & \\ & & \downarrow{\scriptstyle \epsilon} & & \downarrow{\scriptstyle \zeta} & & \\ & & X & \xrightarrow[\theta]{} & Y & \xrightarrow[\varphi]{} & Z , \end{array}$$

where $\beta\alpha$ and $\varphi\theta$ are null-homotopic and $\zeta\beta$ is homotopic to $\theta\epsilon$. When we choose homotopies

$$H \text{ from } * \text{ to } \beta\alpha ,$$
$$J \text{ from } \zeta\beta \text{ to } \theta\epsilon ,$$
$$K \text{ from } * \text{ to } \varphi\theta ,$$

we can form the sequences with homotopy

$$(\theta,\ \epsilon\alpha,\ \{\zeta H,\ J\alpha\}) \doteq s_1 ,$$
$$\text{and } (\varphi\zeta,\ \beta,\ \{K\epsilon,\ \varphi J_\tau\}) \doteq s_2 .$$

The sequence with homotopy s_1 can be combined with (φ, θ, K) to produce the secondary composition

$$\tilde{\varphi} \circ \overline{\epsilon\alpha} \,.$$

The sequence s_2 can be combined with (β, α, H) to produce the secondary composition

$$\underline{\varphi\zeta} \circ \underset{\sim}{\alpha} \,.$$

Thus we can expand the initial array of spaces and maps to

$$\begin{array}{ccccccc}
 & & & & & & \Sigma A \\
 & & & & & & \Big\downarrow \scriptstyle{\underset{\sim}{\alpha}} \\
A & \xrightarrow{\alpha} & B & \xrightarrow{\beta} & C & \xrightarrow{j} & T_\beta \\
\scriptstyle{\overline{\epsilon\alpha}} \Big\downarrow & & \scriptstyle{\epsilon} \Big\downarrow & & \Big\downarrow \scriptstyle{\zeta} & & \Big\downarrow \scriptstyle{\underline{\varphi\zeta}} \\
W_\theta & \xrightarrow[p]{} & X & \xrightarrow[\theta]{} & Y & \xrightarrow[\varphi]{} & Z \\
\scriptstyle{\tilde{\varphi}} \Big\downarrow & & & & & & \\
\Omega Z & & & & & &
\end{array}$$

3.4.2. Basic Peterson-Stein formula. *The adjoint of* $\underline{\varphi\zeta} \circ \underset{\sim}{\alpha}$ *is homotopic to* $\tilde{\varphi} \circ \overline{\epsilon\alpha}$*, up to sign, namely,*

$$(\underline{\varphi\zeta} \circ \underset{\sim}{\alpha})^\natural \simeq (\tilde{\varphi} \circ \overline{\epsilon\alpha})_\tau \,.$$

Proof. We apply the interface and slide Lemmas 3.2.1, 3.2.2. Thus

$$(\underline{\varphi\zeta} \circ \underset{\sim}{\alpha})^\natural \simeq (\widetilde{\varphi\zeta} \circ \bar{\alpha})_\tau \simeq (\tilde{\varphi} \circ \overline{\epsilon\alpha})_\tau \,.$$

□

Next we codify three variations on the basic formula, where either of the following sequences are used:

- **(i):** the cofibration sequence $A \xrightarrow{\alpha} B \xrightarrow{j} T_\alpha \xrightarrow{q} \Sigma A$ where j is the inclusion and q is obtained by pinching B to a point in T_α, or
- **(ii):** the fibration sequence $\Omega Y \xrightarrow{j} W_\theta \xrightarrow{p} X \xrightarrow{\theta} Y$ where j is the inclusion of the fiber.

Corollary 3.4.1. With (i) we have $(\tilde{\varphi}\circ\overline{\epsilon\alpha})^\natural \simeq \underline{\varphi\zeta}\circ\underset{\sim}{\alpha}$ (no direction reversal map) in the homotopy commutative diagram

$$\begin{array}{ccccccc}
A & \xrightarrow{\alpha} & B & \xrightarrow{j} & T_\alpha & \xrightarrow{q} & \Sigma A \\
{\scriptstyle \overline{\epsilon\alpha}}\downarrow & & {\scriptstyle \epsilon}\downarrow & & \downarrow{\scriptstyle \zeta} & & \downarrow{\scriptstyle \underline{\varphi\zeta}\circ\underset{\sim}{\alpha}} \\
W_\theta & \xrightarrow[p]{} & X & \xrightarrow[\theta]{} & Y & \xrightarrow[\varphi]{} & Z \\
{\scriptstyle \tilde{\varphi}}\downarrow & & & & & & \\
\Omega Z & & & & & &
\end{array}$$

Corollary 3.4.2. With (ii), we have $(\tilde{\theta}\circ\overline{\epsilon\alpha})^\natural \simeq \underline{\theta\zeta}\circ\underset{\sim}{\alpha}$ (no direction reversal map) in the homotopy commutative diagram

$$\begin{array}{ccccccc}
 & & & & & & \Sigma A \\
 & & & & & & \downarrow{\scriptstyle \underset{\sim}{\alpha}} \\
A & \xrightarrow{\alpha} & B & \xrightarrow[\beta]{} & C & \xrightarrow{j} & T_\beta \\
{\scriptstyle \tilde{\theta}\circ\overline{\epsilon\alpha}}\downarrow & & {\scriptstyle \epsilon}\downarrow & & \downarrow{\scriptstyle \zeta} & & \downarrow{\scriptstyle \underline{\theta\zeta}} \\
\Omega Y & \xrightarrow[j]{} & W_\theta & \xrightarrow[p]{} & X & \xrightarrow[\theta]{} & Y\ .
\end{array}$$

Corollary 3.4.3. With both (i) and (ii), we have $(\tilde{\theta}\circ\overline{\epsilon\alpha})^\natural \simeq (\underline{\theta\zeta}\circ\underset{\sim}{\alpha})_\tau$ (with the direction reversal map) in the homotopy commutative diagram

$$\begin{array}{ccccccc}
A & \xrightarrow{\alpha} & B & \xrightarrow{j} & T_\alpha & \xrightarrow{q} & \Sigma A \\
{\scriptstyle \tilde{\theta}\circ\overline{\epsilon\alpha}}\downarrow & & \downarrow{\scriptstyle \epsilon} & & \downarrow{\scriptstyle \zeta} & & \downarrow{\scriptstyle \underline{\theta\zeta}\circ\underset{\sim}{\alpha}} \\
\Omega Y & \xrightarrow[j]{} & W_\theta & \xrightarrow[p]{} & X & \xrightarrow[\theta]{} & Y\ .
\end{array}$$

We prove the first corollary. The basic formula puts the sequence

$$A \xrightarrow{\alpha} B \xrightarrow{j} T_\alpha \xrightarrow{\hat{j}} T_j$$

on the top row. From subsection (3.2.7) we have

$$\hat{j} \simeq \underset{\sim_\tau}{\alpha} \circ q$$

and the direction reversal maps cancel, up to homotopy.

The other corollaries follow in a similar manner. When variations other than those codified above appear, we work directly from the basic formula under subheading (3.4.2).

The Corollary 3.4.3 is often applied to the data

$$A \xrightarrow{\alpha} B \xrightarrow{f} X \xrightarrow{\theta} Y \text{ with } f \circ \alpha \sim * \text{ and } \theta \circ f \sim * .$$

Let (f, α, H) and (θ, f, L) be sequences with homotopy. The roles of ϵ and ζ in (3.4.3) are filled by the lifting $\bar{f}$ using L and the extension $\underline{f}$ using H. The result is the following homotopy commutative diagram with the maps given by the extension and lifting constructions (3.2.2), (3.2.4) and the difference construction in (3.3.1),

$$\begin{array}{ccccccc}
A & \xrightarrow{\alpha} & B & \xrightarrow{j_c} & T_\alpha & \xrightarrow{q} & \Sigma A \\
\Big\downarrow{\scriptstyle \delta(L\alpha, \theta H)} & & \Big\downarrow{\scriptstyle \bar{f}} & & \Big\downarrow{\scriptstyle \underline{f}} & & \Big\downarrow{\scriptstyle \delta(\theta H, L\alpha)} \\
\Omega Y & \xrightarrow[j]{} & W_\theta & \xrightarrow[p]{} & X & \xrightarrow[\theta]{} & Y
\end{array} .$$

In particular, when the map j is substituted for the map α, we have

$$\begin{array}{ccccccc}
\Omega Y & \xrightarrow{j} & W_\theta & \longrightarrow & T_j & \xrightarrow{q} & \Sigma \Omega Y \\
\Big\| & & \Big\| & & \Big\downarrow{\scriptstyle \underline{p}} & & \Big\downarrow{\scriptstyle -(\mathrm{eval})} \\
\Omega Y & \xrightarrow{j} & W_\theta & \xrightarrow[p]{} & X & \xrightarrow[\theta]{} & Y
\end{array} .$$

3.4.3. A calculation. *The mod 2 Hurewicz map for Spin*(2^m) *is non-zero in dimension* $2^m - 1$, $m \geq 2$,

$$h : \pi_{2^m-1}(Spin(2^m)) \to H_{2^m-1}(Spin(2^m); Z/2) .$$

Of course, this is of interest only for $m \geq 4$ since Spin(4) $= S^3 \times S^3$, Spin(8) $=$ Spin(7) $\times S^7$.

To prove this, we first recall some facts about the mod 2 cohomology of $SO(n)$, [**109**] or [**117**]. The real projective space RP^{n-1} can be embedded in $SO(n)$ and the embedding

$$\epsilon : RP^{n-1} \to SO(n)$$

induces an isomorphism on mod 2 cohomology in dimension 1. Writing the generators as

$$u \text{ in } H^1(SO(n); Z/2) ,$$
$$x \text{ in } H^1(RP^{n-1}; Z/2) ,$$

we have $\epsilon^*(u) = x$ and $u^{2^n} = 0$. Thus, we can set up the array of spaces and maps, as in Cor. 3.4.1.

$$
\begin{array}{ccccccc}
S^{2^m-1} & \xrightarrow{\alpha} & RP^{2^m-1} & \xrightarrow{j} & RP^{2^m} & \xrightarrow{q} & S^{2^m} \\
 & & \downarrow{\scriptstyle \epsilon} & & \downarrow{\scriptstyle \zeta} & & \\
\mathrm{Spin}(2^m) & \xrightarrow[p]{} & SO(2^m) & \xrightarrow[\theta]{} & K(Z/2,1) & \xrightarrow[\varphi]{} & K(Z/2,2^m)
\end{array}
$$

where the row is a cofibration sequence,

$$
\begin{aligned}
\theta^*(b_1) &= u \,, \\
\varphi^*(b_{2^m}) &= Sq(2^{m-1}, \ldots, 1)b_1 \,, \\
\zeta^*(b_1) &= x \,,
\end{aligned}
$$

and we regard $\mathrm{Spin}(2^m)$ as the homotopy theory fiber of θ. Since

$$\zeta^* \varphi^* b_{2^m} = x^{2^m} \neq 0$$

in

$$H^*(RP^{2^m}; Z/2)$$

and q^* is an isomorphism in dimension 2^m with $Z/2$ coefficients, there is a map on the left with non-zero Hurewicz image.

Our Peterson-Stein formula is an adjoint (up to sign) relationship between certain secondary compositions. We consider now what results from the set-up in subsection (3.4.1), but where arbitrary maps fill in the squares, up to homotopy. Consider the homotopy commutative diagram

$$
\begin{array}{ccccccc}
 & & & & & & \Sigma A \\
 & & & & & & \downarrow{\scriptstyle \underset{\sim}{\alpha}} \\
A & \xrightarrow{\alpha} & B & \xrightarrow{\beta} & C & \longrightarrow & T_\beta \\
\downarrow{\scriptstyle \epsilon'} & & \downarrow{\scriptstyle \epsilon} & & \downarrow{\scriptstyle \zeta} & & \downarrow{\scriptstyle \zeta'} \\
W_\theta & \xrightarrow[p]{} & X & \xrightarrow[\theta]{} & Y & \xrightarrow[\varphi]{} & Z \\
\downarrow{\scriptstyle \tilde{\varphi}} & & & & & & \\
\Omega Z & & & & & &
\end{array}
$$

where $\underset{\sim}{\alpha}$ is the coextension from (β, α, H) and $\tilde{\varphi}$ is the colifting from $(\varphi, \theta, \kappa)$ but ϵ' and ζ' are assumed only to make the squares commute up to homotopy.

Then we have

$$[\![\tilde{\varphi}\circ\epsilon']\!] = [\![\tilde{\varphi}\circ\overline{\epsilon\alpha}]\!] \text{ in } [A,\Omega Z]/im\Omega\varphi_{\#}$$
$$\text{and } [\![\zeta'\circ\underset{\sim}{\alpha}]\!] = [\![\underline{\varphi\zeta}\circ\underset{\sim}{\alpha}]\!] \text{ in } im\Sigma\alpha\# \setminus [\Sigma A, Z]\ .$$

Hence (we assume, for convenience that Z is a simply connected H-space)

$$[\![(\tilde{\varphi}\circ\epsilon')^{\natural}]\!] = [\![(\zeta'\circ\underset{\sim}{\alpha})_{\tau}]\!] \text{ in the quotient}$$
$$[\Sigma A, Z]/im\varphi_{\#} + im\Sigma\alpha^{\#}\ .$$

Moreover, it follows that the equation is independent of the homotopies used to construct the coextension or colifting.

The facts latent in the set-up of (3.4.1) are the values of either $\varphi\zeta$ or $\epsilon\alpha$. The calculation made in subsection (3.4.3) is an illustration of the (infrequent) situation where a conclusion of interest may be obtained without dividing. More often, conclusions based on these facts must be drawn after passage to some quotient involving $\Omega\varphi_{\#}$ or $\Sigma\alpha^{\#}$.

3.5. The Milnor filtration

In this section, we discuss a filtration which provides a way to tell how a null-homotopic composition of loop maps

$$\Omega X \xrightarrow{\Omega\theta} \Omega Y \xrightarrow{\Omega\varphi} \Omega Z$$

got that way. The filtration on X will be used to factor the composite $\varphi\theta$ through some k-fold smash product, suggested in the following diagram

$$\begin{array}{ccccc} B_k & \longrightarrow & B_{k+1} & \longrightarrow & (\Sigma\Omega X)^{(k)} \\ & & \downarrow & & \downarrow \\ & & X & \xrightarrow[\varphi\theta]{} & X\ , \end{array}$$

where the spaces B_k form the filtration of X, and the top row is a cofibration sequence.

3.5.1. Filtering a space with the Dold-Lashof construction. Our filtration is made with the Dold-Lashof construction. We leave the pointed category $\mathcal{K}_*$ for the moment, but continue to work in $\mathcal{K}$, with compactly generated spaces. Basepoints are present and most of the maps will preserve them. But we do not reduce cones or suspensions, and homotopies are not required to preserve basepoints.

There are several special features which we discuss first of all. We refer to Stasheff's book [**108**] for details and references to the literature, as well as [**28**], [**19**] .

3.5.2. Measured loops (a.k.a. Moore loops). Let X be a space with basepoint $*$. We write $\Omega^* X$ for the space of *measured loops* as discussed in [**117**]. We recall that the H-structure

$$\Omega^* X \times \Omega^* X \xrightarrow{+} \Omega^* X \text{ is given by}$$
$$(\omega_1 + \omega_2) : [0, \gamma_1 + \gamma_2] \to X \ ,$$
$$(\omega_1 + \omega_2)(t) = \begin{Bmatrix} \omega_1(t), & 0 \le t \le \gamma_1 \\ \omega_2(t - \gamma_1), & \gamma_1 \le t \le \gamma_1 + \gamma_2 \end{Bmatrix}$$

where

$$\omega_i : [0, \gamma_i] \to X \text{ satisfies}$$
$$\omega_i(0) = * = \omega_i(\gamma_i) \quad i = 1, 2, \ .$$

The identity for addition is denoted by e, and is the map sending 0 to the basepoint of X.

3.5.3. Quasi-fibrations. A map $f : X \to Y$ is called a *quasi-fibration* provided that it is surjective and induces an isomorphism on homotopy groups

$$\pi_*(X, f^{-1}(y)) \to \pi_*(Y, y)$$

for each y in Y. Quasi-fibrations arise from

3.5.4. The Hopf construction. Given a surjection

$$f : X \times Y \to Z$$

the classical Hopf construction

$$H(f) : X * Y \to \Sigma Z \quad \text{(unreduced suspension)}$$

is given, in coordinates, by the map sending (x, t, y) to $(t, f(x, y))$. Here

$$X * Y = X \times I \times Y \big/ R$$

where R is the relation generated by

$$(x, 0, y) \sim (x', 0, y) \ ,$$
$$(x, 1, y) \sim (x, 1, y') \ .$$

When $X = Z$, we rework this construction, with notation anticipating the Dold-Lashof construction. Write

$$DL(X) = X \cup_f (CX \times Y)$$

for the adjunction space formed by attaching

$$CX \times Y \quad \text{(unreduced cone on } X \text{ with vertex at 0)}$$

to X along the subspace $\{1\} \times X \times Y$ by f.

If X and Y have the homotopy type of CW complexes and the *shearing map*

$$S : X \times Y \to X \times Y \text{ given by } S(x, y) = (f(x, y), y)$$

is a homotopy equivalence, then there is a homotopy equivalence

$$DL(X) \simeq X * Y .$$

To see this, embed X in $X \times CY$ by sending the point x to $(x, 1, *)$ where $*$ is the basepoint for Y. Composition of this embedding with f yields a homotopy equivalence,

$$DL(X) \subset DL(X \times CY) .$$

Then S together with the homotopy contracting Y to $(1, *)$ in CY provides a homotopy between the adjunction maps for

$$DL(X \times CY) \text{ and } X * Y = X \times CY \bigcup_{X \times Y} CX \times Y .$$

Given a surjection

$$p : X \to B$$

such that $pf(x, y) = p(x)$, then there is an induced map

$$DL(p) : DL(X) \to B \cup_p CX$$

determined by projecting $CX \times Y$ to CX. In the case where B is a point, we have recovered the Hopf construction, up to homotopy

$$\begin{array}{ccc} DL(X) & \simeq & X * Y \\ & {\scriptstyle DL(p)} \searrow \quad \swarrow {\scriptstyle H(f)} & \\ & \Sigma X & , \end{array}$$

and more importantly, $DL(p)$ *is a quasi-fibration.*

For the case of a connected H-space of the homotopy type of a CW complex, then the shearing map for the multiplication induces an isomorphism of homotopy groups, and hence, a homotopy equivalence.

For B in general, the map

$$DL(p) : DL(X) \to DL(B) = B \cup_p CX$$

is a quasi-fibration provided that for each point x in X, the map

$$Y \to p^{-1}p(x) ,$$

given by sending y to $f(x, y)$, is a weak equivalence.

To summarize, starting with

$$p : X \to B$$

and a *fiberwise action* (i.e. $pf(x,y) = p(x)$)

$$f : X \times Y \to X$$

a weak equivalence of "fibers"; that is, the map

$$Y \to p^{-1}p(x) \text{ where } y \mapsto f(x,y)$$

is a weak equivalence for each point x in X, then we receive a quasi-fibration

$$DL(p) : DL(X) \to DL(B) .$$

This construction is known as the *Dold-Lashof construction* on p using f.

3.5.5. Iteration of the Dold-Lashof construction. When Y is an associative H-space, the construction in (3.5.4) can be iterated. We want a fiberwise action,

$$\mu : DL(X) \times Y \to DL(X) ,$$

over $DL(p)$, which induces a weak equivalence of fibers. At the point level, the formula for μ is easy to write,

$$\mu((t,x,y_1),y_2) = (t,x,y_1 + y_2)$$

where $+$ denotes the H-structure on Y. The fact that μ is continuous is a convenient feature of K, [**108**]. In particular, Y has been given the associated compactly generated topology, and $DL(X)$ is constructed from a proclusion

$$X \amalg CX \times Y \to DL(X) .$$

Moreover, we work in the setting where the Hausdorff condition in [**117**] has been relaxed to weak Hausdorff (the diagonal is closed in the product (in K)). A modern treatment of the Dold-Lashof construction in this setting is included in [**88**].

The product in $\mathcal{K}$ preserves proclusions, hence the source of μ is an identification space.

If we fix a point in $DL(X)$, say (t,x,y_0), then μ induces

$$Y \to \{t,x\} \times Y$$

given by

$$y \to (\{t,x\}, y_0 + y) ,$$

which is a weak equivalence, (we use this material only for connected spaces, so we assume Y is connected even though the theory works more generally). Thus we have the ingredients to repeat the Dold-Lashof construction on $DL(p)$ using μ.

Starting with a simply connected space Y and iterating the Dold-Lashof construction from the initial data

$$\Omega^* Y \to *$$

leads to

Proposition 3.5.1. Let Y be a pointed simply connected space with the homotopy type of a CW complex. There exist spaces B_kY, $k \geq 0$, with the following properties:

(a): $B_0Y = *$ and $B_kY \subset B_{k+1}Y$. We write $j_{k,1}$ for the inclusion and $j_{k,i}$ for the i-th composition of inclusions,

$$j_{k,i} : B_kY \to B_{k+i}Y \ .$$

(b): Given a base point preserving map $f : Y \to Z$, there are maps

$$B_kf : B_kY \to B_kZ$$

such that $B_{k+i}f \circ j_{k,i} = j_{k,i} \circ B_kf$.

(c): There is a homotopy equivalence

$$\phi_Y : \bigcup_{k \geq 0} B_kY \to Y \ .$$

Write $j_{k,\infty}$ for the restriction of ϕ_Y to B_kY. Then

$$j_{k,\infty} \circ B_kf = f \circ j_{k,\infty} \ .$$

(d): The space B_1Y is $\Sigma\Omega^*Y$, $B_1f = \Sigma\Omega^*f$ and $j_{1,\infty}$ is the evaluation map.

(e): There are spaces $E_k\Omega Y, k \geq 0$, and quasi-fibrations

$$p_k : E_k\Omega Y \to B_kY$$

such that $B_{k+1}Y$ is the (unreduced) mapping cone of p_k. For $k = 0$, $E_0\Omega Y = \Omega^*Y$ and p_0 is the constant map. For $k = 1$, we can identify $p_1 : E_1\Omega Y \to B_1Y$ with the Hopf construction. Furthermore, there are homotopy equivalences

$$\Sigma E_k\Omega Y \simeq (\Sigma\Omega Y)^{(k+1)}$$

of $\Sigma E_k\Omega Y$ with the $(k+1)$-fold smash product.

(f): There is a relative homeomorphism

$$(CE_k\Omega Y, E_k\Omega Y) \times \Omega^*Y \to (E_{k+1}\Omega Y, E_k\Omega Y) \ .$$

Everything, except (c) comes from iterating the Dold-Lashof construction. Thus

$$E_{k+1}\Omega Y \xrightarrow{p_{k+1}} B_{k+1}Y$$

is just

$$DL(E_k\Omega Y) \xrightarrow{DL(p_k)} DL(B_kY) = \text{ mapping cone of } p_k \ .$$

The map $j_{1,\infty}$ is also easily described. For $k = 1$

$$B_1Y = \Sigma\Omega^*Y \text{ by construction.}$$

Define $j_{1,\infty}$ by

$$j_{1,\infty}(s,\omega) = \omega(rs)$$

where $\omega : [0,r] \to$Y. For (c), we refer to [**108**], p. 17–18.

Remark. J. Moore points out that the Princeton thesis by H. Hastings develops this theory for simplicial monoids Γ. Then the Eilenberg-Mac Lane filtration on $\bar{W}(\Gamma)$ gives the Milnor filtration on the geometric realization, for the singular complex of the Moore loop space.

3.5.6. Cohomology operations and the Milnor filtration. We intend to use the cofibration sequences from Prop. 3.5.1(e) as top rows in the arrays discussed in subsection (3.4.1). There are two elementary points to consider, returning to $\mathcal{K}_*$ and signs.

With respect to base points, it is necessary only to reduce suspensions, e.g. $\Sigma E_k \Omega Y$, but not the constructions described in Prop. 3.5.1. The base point e in $\Omega^* Y$ is preserved by $j_{k,i}$ and is mapped to the base point of Y by $j_{k,\infty}$. Likewise for

$$p_k : E_k \Omega Y \to B_k Y\ .$$

Since the composition

$$j_{k,1} \circ p_k$$

is null-homotopic and $B_{k+1}Y$ is simply connected, we have contracting homotopies in $\mathcal{K}_*$. The usual loop space ΩY is contained in $\Omega^* Y$ as a deformation retract, and the diagram

$$\begin{array}{ccc}
\Sigma\Omega Y & \xrightarrow{\Sigma(\text{inc})} & \Sigma\Omega^* Y \quad \text{(unreduced suspension)} \\
\underline{\text{reduce}}\big\downarrow & & \big\downarrow j_{1,\infty} \\
\Sigma\Omega Y & \xrightarrow{\quad p \quad} & Y \quad \text{(reduced suspension)}
\end{array}$$

is commutative, where p is the usual evaluation.

With respect to signs, we proceed deliberately, because the Milnor filtration leads to an array of the type introduced in (3.4.1), but not codified there. We state the result. Let

$$q : B_{k+1}Y \to \Sigma E_k \Omega Y$$

be obtained by pinching $B_k Y$ to a point in $B_{k+1}Y$ and then reducing in the suspension coordinate.

Proposition 3.5.2. Starting with the array, where $\zeta q \simeq \theta\epsilon$,

$$\begin{array}{ccccc} B_kY & \xrightarrow{j_{k,1}} & B_{k+1}Y & \xrightarrow{q} & \Sigma E_k \Omega Y & & \\ & & \downarrow{\scriptstyle \epsilon} & & \downarrow{\scriptstyle \zeta} & & \\ & & C_0 & \xrightarrow[\theta]{} & C_1 & \xrightarrow[\varphi]{} & C_2 \end{array}$$

and a contracting homotopy H for $\varphi\theta$. Then there is a homotopy commutative diagram

$$\begin{array}{ccccccc} B_kY & \xrightarrow{j_{k,1}} & B_{k+1}Y & \xrightarrow{q} & \Sigma E_k \Omega Y & \xrightarrow{\Sigma p_k} & \Sigma B_k Y \\ {\scriptstyle \eta}\downarrow & & \downarrow{\scriptstyle \epsilon} & & \downarrow{\scriptstyle \zeta} & & \downarrow{\scriptstyle (\tilde{\varphi}\circ\eta)^\flat_\tau} \\ W_\theta & \xrightarrow[p]{} & C_0 & \xrightarrow[\theta]{} & C_1 & \xrightarrow[\varphi]{} & C_2 \\ {\scriptstyle \tilde{\varphi}}\downarrow & & & & & & \\ \Omega C_2 & & & & & & \end{array}$$

where $\tilde{\varphi}$ is the colifting associated to (φ, θ, H); and, as indicated in the diagram, the outer maps are adjoint up to sign.

Proof. We combine (3.4.2) with the sign information in (3.2.7) to form the following diagram,

$$\begin{array}{ccccccc} & & B_{k+1} & \xrightarrow[\boxed{-1}]{q} & \Sigma E_k \Omega Y & \xrightarrow[\boxed{-1}]{\Sigma p_k} & \Sigma B_k Y \\ & & \| & & \downarrow{\scriptstyle \underset{\sim}{p}_k} & & \downarrow{\scriptstyle \underset{\sim}{j}_{k,1}} \\ B_kY & \xrightarrow{j_{k,1}} & B_{k+1}Y & \xrightarrow{\hat{j}_{k,1}} & T_{j_{k,1}} & \xrightarrow{h} & T_{\hat{j}_{k,1}} \\ {\scriptstyle \overline{\epsilon\circ j_{k,1}}}\downarrow & & \downarrow{\scriptstyle \epsilon} & & \downarrow{\scriptstyle \zeta'} & & \downarrow{\scriptstyle \underline{\varphi\zeta'}} \\ W_\theta & \xrightarrow[p]{} & C_0 & \xrightarrow[\theta]{} & C_1 & \xrightarrow[\varphi]{} & C_2 \\ {\scriptstyle \tilde{\varphi}}\downarrow & & & & & & \\ \Omega C_2\ , & & & & & & \end{array}$$

where the two top squares anti-commute. The map ζ' is chosen to satisfy

$$\zeta' \circ ((-1) \text{ times } \underset{\sim}{p_k}) \simeq \zeta\ ,$$

(which is possible because $\underset{\sim}{p_k}$ is a homotopy equivalence). The outer maps are constructed from fixed choices of a contracting homotopy for $\hat{j}_{k,1} \circ j_{k,1}$ and a homotopy from $\zeta' \circ \hat{j}_{k,1}$ to $\theta \circ \epsilon$, along with H. Furthermore, the names of the maps are those from Prop. 3.5.1 where there is notational agreement. From (3.4.2) we have

$$\left(\tilde{\varphi} \circ \overline{\epsilon j_{k,1}}\right)^{\flat} \simeq (-1) \text{ times } \underline{\varphi\zeta'} \circ \underset{\sim}{j_{k,1}} \,.$$

Then, composing the right side with Σp_k, and observing that the two (-1) signs cancel yields the stated result. □

The next result is about the value of ζ in Prop. 3.5.2 for certain values of θ. For this we write

$$(\Sigma\Omega Y)^{(k)} \quad k - \text{fold smash}$$

in place of

$$\Sigma E_k \Omega Y$$

but continue to use the letter q to name the map.

Proposition 3.5.3. We have the following homotopy commutative diagrams:

(a): for $Y = K(Z/2, n)$, $n \geq 2$,

$$\begin{array}{ccc} B_2 Y & \xrightarrow{\;q\;} & (\Sigma\Omega Y)^{(2)} \\ {\scriptstyle j_{2,\infty}}\downarrow & & \downarrow{\scriptstyle [b_{n-1}|b_{n-1}]} \\ Y & \xrightarrow[Sq^n]{} & K(Z/2, 2n) \end{array}$$

(b): for $Y = K(Z/p, 2n+1)$, $n \geq 1$, p an odd prime,

$$\begin{array}{ccc} B_2 Y & \xrightarrow{\;q\;} & (\Sigma\Omega Y)^{(2)} \\ {\scriptstyle j_{2,\infty}}\downarrow & & \downarrow{\scriptstyle -\sum_{i=1}^{p-1} \frac{1}{p}\binom{p}{i}[b_{2n}^i | b_{2n}^{p-i}]} \\ Y & \xrightarrow[\boldsymbol{\beta}\mathcal{P}^n]{} & K(Z/p, 2np+2) \end{array}$$

(signed Bockstein)

(c): for $Y = K(Z/p, 2n)$, $n \geq 1$, p an odd prime,

$$\begin{array}{ccc} B_p Y & \xrightarrow{\;q\;} & (\Sigma\Omega Y)^{(p)} \\ {\scriptstyle j_{p,\infty}}\downarrow & & \downarrow{\scriptstyle [b_{2n-1}|\cdots|b_{2n-1}]} \\ Y & \xrightarrow[\mathcal{P}^n]{} & K(Z/p, 2np)\,. \end{array}$$

The values for ζ are given using the notation for the bar construction on $H^*(\Omega Y)$.

Proof. Some such diagrams exist because

$$Sq^n \circ j_{1,\infty} \text{ and } \beta\mathcal{P}^n \circ j_{1,\infty}$$

are both null-homotopic by the zero property; and

$$\mathcal{P}^n \circ j_{p-1,\infty}$$

represents

$$j^*_{p-1,\infty}(b^p_{2n}) = 0$$

since $B_{p-1}Y$ is a space with Lusternik-Schnirelmann category at most $(p-1)$. So the question is to identify ζ.

For (a) we know

$$H^{2n}(\Sigma\Omega Y^{(2)}) = Z/2$$

with generator $[b_{n-1} \mid b_{n-1}]$. The cohomology of the "projective plane" B_2Y has the cup product

$$j^*_{2,\infty}(b_n) \cup j^*_{2,\infty}(b_n) = q^*[b_{n-1} \mid b_{n-1}],$$

and the left side is represented by

$$Sq^n \circ j_{2,\infty} .$$

The cup product formula above is proved in [**16**]. Observe, also, that it holds for dimensional reasons, given the information that the cup square of b_n is non-zero.

For (b) and (c), we make use of information supplied by Cartan exposés 6–8, 16, [**20**]. Let A_* denote the homology Hopf algebra $H_*(\Omega Y)$. We know that p-th powers in A are 0, so the transpotence operator, from A_* to the submodule of primitives,

$$\varphi_p : A_* \longrightarrow PH_*(Y)$$

is defined. Let c be the homology class of dimension $2n$,

$$c \text{ in } H_{2n}(\Omega Y) \text{ such that } \langle c, b_{2n}\rangle = 1 .$$

We calculate $\varphi_p(c)$ using the bar construction

$$(A_*, \bar{B}(A_*), B(A_*)) .$$

By the definition of φ_p in [**20**], a representative cycle for $\varphi_p(c)$ is

$$[c^{p-1}|c] \text{ in } \bar{B}(A_*) .$$

Cartan proves in exposé 16 of [**20**], that the evaluation pairing satisfies

$$\langle \varphi_p(c), \beta\mathcal{P}^n b_{2n+1}\rangle = 1 .$$

We know that

$$\zeta = \frac{1}{p}\sum_{i=1}^{p-1}\binom{p}{i}[b^i_{2n} \mid b^{p-i}_{2n}] \text{ in } \bar{B}(A^*)$$

is a cycle which represents an element in

$$PH^{2np+2}(Y) .$$

Furthermore, the dimension of

$$PH^{2np+2}(Y) \cap \ker \sigma^*$$

is 1, so ζ represents $\beta\mathcal{P}^n$ up to some coefficient in Z/p. Evaluating

$$\langle [c^{p-1} \mid c], \zeta \rangle \equiv (p-1)! \equiv -1 \mod p$$

gives (b).

For (c), the argument is similar, using Cartan's results on divided powers in $\bar{B}(A_*)$, for

$$A_* = H_*(K(Z/p, 2n-1)) .$$

Here, the numerical information is

$$\gamma_p[c] = [c|\cdots|c] \text{ in } \bar{B}_p(A_*) ,$$
$$\langle \gamma_p[c], \mathcal{P}^n b_{2n} \rangle = 1 ,$$

and

$$\dim PH^{2np}(Y) \cap \ker \sigma^* = 1 .$$

□

3.5.7. Maps which loop to 0. Given $f : Y \to Z$ with Ωf null-homotopic, then ΩW_f splits as a space, up to homotopy, as $\Omega Y \times \Omega^2 Z$. We work out some details in terms of the Milnor filtration.

Let $k \geq 1$ be an integer such that the composition

$$B_k Y \xrightarrow{j_{k,\infty}} Y \xrightarrow{f} Z$$

is null-homotopic. In practice, we are interested in the maximal value of k, but do not impose that condition here. Choosing a homotopy for the sequence

$$(f \circ j_{k+1,\infty}, j_{k,1}, L)$$

gives an extension

$$\begin{array}{ccc} B_{k+1}Y & \xrightarrow{\hat{j}} & T \\ {\scriptstyle j_{k+1,\infty}}\downarrow & & \downarrow{\scriptstyle \hat{f}} \\ Y & \xrightarrow[f]{} & Z \end{array}$$

where T is the mapping cone of $j_{k,1}$ and $\hat{f}$ is the extension

$$\hat{f} = \underline{f \circ j_{k+1,\infty}} .$$

We can write

$$\hat{f} \circ \underset{\sim}{p}_k \simeq (-1) \text{ times } \zeta$$

in the diagram

$$\begin{array}{ccc} B_{k+1}Y & \xrightarrow[\boxed{-1}]{q} & (\Sigma\Omega Y)^{(k+1)} \\ \| & & \downarrow \underset{\sim}{p}_k \\ B_{k+1}Y & \xrightarrow[\hat{j}]{} & T \\ \downarrow & & \downarrow \hat{f} \\ Y & \xrightarrow{f} & Z \end{array}$$

the top square anti-commutes.

We pause, to indicate the role of ζ for the discussion of ΩW_f. In the sequel we shall construct a certain map

$$\gamma^{\natural\natural} : \Omega W_f \to \Omega^2 Z$$

which encodes the null-homotopy for Ωf. Among other things, we have a homotopy equivalence

$$\Omega p \times \gamma^{\natural\natural} : \Omega W_f \to \Omega Y \times \Omega^2 Z \; .$$

In the case where $k = 1$, the H-deviation of $\gamma^{\natural\natural}$ is measured by ζ, and otherwise we have an H-equivalence.

From the homotopy fiber of f

$$\begin{array}{ccc} W_f & \xrightarrow{F} & P_0Z \\ p\downarrow & & \downarrow \\ Y & \xrightarrow[f]{} & Z \end{array}$$

and the observation

$$f \circ p \circ j_{k,\infty} = f \circ j_{k,\infty} \circ B_k(p)$$

we obtain a second null-homotopy

$$F^{\flat} \circ T_0 j_{k,\infty} \; .$$

We measure the difference in terms of a map from $\Sigma B_k W_f$ to Z. There are two reasonable choices. We define

$$\gamma : \Sigma B_k W_f \to Z$$

by means of the push-out diagram

$$\begin{array}{ccccc} T_1B_kW_f & \longleftarrow & B_kW_f & \longrightarrow & T_0B_kW_f \\ \Big\downarrow{\scriptstyle L_T\circ T_1B_k(p)} & & \Big\downarrow{\scriptstyle f\circ p\circ j_{k,\infty}} & & \Big\downarrow{\scriptstyle F^\flat\circ T_0 j_{k,\infty}} \\ Z & = & Z & = & Z\,. \end{array}$$

It follows that γ is *compatible with the evaluation map.* The diagram

$$\begin{array}{ccc} \Sigma B_1\Omega Z & \simeq & \Sigma^2\Omega^2 Z \\ {\scriptstyle \Sigma(j_{1,k}\circ B_1(\mathrm{inc}))}\Big\downarrow & & \Big\downarrow{\scriptstyle \mathrm{eval.}} \\ \Sigma B_kW_f & \underset{\gamma}{\longrightarrow} & Z \end{array}$$

commutes up to homotopy, where (inc) is the inclusion or ΩZ in W_f. To see this, we have

$$B_k(p)\circ j_{1,k}\circ\ (\mathrm{inc})\ = * \ ,$$

and L is a pointed homotopy. Hence, up to homotopy, the composite

$$\gamma\circ\Sigma(j_{1,k}\circ B_1\ (\mathrm{inc})\ )$$

factors through the map collapsing $T_1B_1\Omega Z$ to a point in $\Sigma B_1\Omega Z$. Moreover, the composition

$$T_0B_1\Omega Z \longrightarrow T_0B_1W_f \longrightarrow T_0W_f \xrightarrow{F^\flat} Z$$

is the evaluation map.

In the next diagram we note that the existence of some map filling in on the right is immediate. However, we want to know that our construction gives such a map. The reason for this demand is in order to identify a suitable double adjoint with a colifting.

Proposition 3.5.4. The following diagram commutes up to homotopy:

$$\begin{array}{ccccc} B_{k+1}W_f & \xrightarrow{q} & (\Sigma\Omega W_f)^{(k+1)} & \xrightarrow{\Sigma p_k} & \Sigma B_kW_f \\ \Big\downarrow & & \Big\downarrow{\scriptstyle (\Sigma\Omega p)^{(k+1)}} & & \Big\downarrow \\ B_{k+1}Y & \longrightarrow & (\Sigma\Omega Y)^{k+1)} & & {\scriptstyle\gamma} \\ \Big\downarrow & & \Big\downarrow{\scriptstyle\varsigma} & & \Big\downarrow \\ Y & \underset{f}{\longrightarrow} & Z & = & Z\ , \end{array}$$

and γ is compatible with the evaluation map.

Proof. We have a commutative diagram, induced by p,

$$\begin{array}{ccccccc}
B_k W_f & \xrightarrow{j_{k,1}} & B_{k+1}W_f & \xrightarrow{\hat{\jmath}'} & T' & \xrightarrow{\hat{f}'} & Z \\
\downarrow & & \downarrow & & \downarrow & & \| \\
B_k Y & \xrightarrow[j_{k,1}]{} & B_{k+1}Y & \xrightarrow[\hat{\jmath}]{} & T & \xrightarrow[\hat{f}]{} & Z
\end{array}$$

and we first show that γ is the secondary composition from the top row, using the canonical homotopy for $\hat{\jmath}' \circ j_{k,1}$ and

$$F^\flat \circ T_0 j_{k+1,\infty} \ .$$

In terms of push-outs, $\hat{f}'$ is determined by

$$\begin{array}{ccccc}
B_{k+1}W_f & \xleftarrow{j_{k,1}} & B_k W_f & \longrightarrow & T_0 B_k W_f \\
\Big\downarrow{\scriptstyle f\circ j_{k+1,\infty}\circ B_k(p)} & & \Big\downarrow{\scriptstyle f\circ j_{k,\infty}\circ B_k(p)} & & \Big\downarrow{\scriptstyle L\circ T_0 B_k(p)} \\
Z & = & Z & = & Z \ .
\end{array}$$

Since the map on the left is $f \circ p \circ j_{k+1,\infty}$, we can use $F^\flat$ to construct an extension

$$\underline{\hat{f}'} : T_{\hat{\jmath}'} \longrightarrow Z \ .$$

Moreover, we can compose this extension with the coextension $\underset{\sim}{j_{k,1}}$. The defining push-out diagrams are juxtaposed,

$$\begin{array}{ccccc}
T_1 B_k W_f & \longleftarrow & B_k W_f & \longrightarrow & T_0 B_k W_f \\
{\scriptstyle (\mathrm{can})}\Big\downarrow & & \Big\downarrow{\scriptstyle j_{k,1}} & & \Big\downarrow \\
T' & \xleftarrow{\hat{\jmath}'} & B_{k+1}W_f & \longrightarrow & T_0 B_{k+1} W_f \\
{\scriptstyle \hat{f}'}\Big\downarrow & & \Big\downarrow{\scriptstyle \hat{f}'\circ\hat{\jmath}'} & & \Big\downarrow{\scriptstyle F^\flat\circ T_0 j_{k+1,\infty}} \\
Z & = & Z & = & Z \ .
\end{array}$$

Since $\hat{f}'$ composed with the canonical null-homotopy for $\hat{\jmath}' \circ j_{k,1}$ is

$$L_\tau \circ T_1 B_k(p) \ ,$$

we have displayed γ as a secondary composition. To obtain the proposition as stated, we observe the diagram

$$
\begin{array}{ccccc}
B_{k+1}W_f & \xrightarrow{q} & E_{k+1}\Omega W_f & \xrightarrow{-\Sigma p_k} & \Sigma B_k W_f \\
\| & \boxed{-1} & \downarrow & & \downarrow {\scriptstyle j_{k,1}} \, \sim \\
B_{k+1}W_f & \longrightarrow & T' & \longrightarrow & T_{\hat{j}'} \\
\downarrow & & \downarrow & & \\
B_{k+1}Y & \longrightarrow & T & & \Big\downarrow {\scriptstyle \underline{\hat{f}'}} \\
\downarrow & & \downarrow {\scriptstyle \hat{f}} & & \\
Y & \xrightarrow{f} & Z & = & Z
\end{array}
$$

where the two minus signs cancel, since the upper left square anti-commutes. □

Finally, we produce the colifting promised at the start of this subsection. We look at the composition

$$\Omega W_f \xrightarrow{j^{\natural}_{1,k}} \Omega B_k W_f \xrightarrow{\Omega(\gamma^{\natural})} \Omega^2 Z$$

with special attention to the case $k = 1$. In terms of adjoints in the diagram defining γ, we have

$$\gamma^{\natural\natural} : \Omega W_f \longrightarrow \Omega^2 Z$$

given by

$$
\begin{array}{ccccc}
\Omega W_f & = & \Omega W_f & = & \Omega W_f \\
{\scriptstyle \Omega(L^{\natural}_{T})\circ\Omega p} \downarrow & & \downarrow {\scriptstyle \Omega(f\circ p)} & & \downarrow {\scriptstyle \Omega F} \\
\Omega P_1 Z & \longrightarrow & \Omega Z & \longleftarrow & \Omega P_0 Z \, .
\end{array}
$$

With the constructions of subsection (3.2.7) for the case $\alpha = f$, $\beta = Id$ we have a colifting

$$\tilde{I} : W_{T\circ\Omega f} \longrightarrow \Omega^2 Z$$

from the diagram

$$\begin{array}{ccccc} \Omega Y & \xrightarrow{\tau\circ\Omega f} & \Omega Z & \longleftarrow & P_0\Omega Z \\ {\scriptstyle (\tau\circ\Omega L_\tau)^\natural}\downarrow & & \| & & \| \\ P_1\Omega Z & \longrightarrow & \Omega Z & \longleftarrow & P_0\Omega Z\ . \end{array}$$

Proposition 3.5.5. The following diagram commutes up to homotopy:

$$\begin{array}{ccc} \Omega W_f & \xrightarrow{\gamma^{\natural\natural}} & \Omega^2 Z \\ {\scriptstyle \rho}\downarrow & & \downarrow{\scriptstyle \tau\circ\kappa} \\ W_{\tau\circ\Omega f} & \xrightarrow[\tilde{I}]{} & \Omega^2 Z\ . \end{array}$$

Proof. The argument is the same as in (3.2.7), inspection of the simpler diagram,

$$\begin{array}{ccccccc} \Omega Y & = & \Omega Y & \xrightarrow{\Omega f} & \Omega Z & \longleftarrow & \Omega P_0 Z \\ \| & & \| & & \downarrow{\scriptstyle \tau} & & \downarrow{\scriptstyle \tau\circ\kappa} \\ \Omega Y & = & \Omega Y & \xrightarrow{\tau\circ\Omega f} & \Omega Z & \longleftarrow & P_0\Omega Z \\ {\scriptstyle \Omega(L_\tau^\natural)}\downarrow & & \downarrow{\scriptstyle (\tau\circ\Omega L_\tau)^\natural} & & \| & & \| \\ \Omega P_1 Z & \xrightarrow[\tau\circ\kappa]{} & P_1\Omega Z & \longrightarrow & \Omega Z & \longleftarrow & P_0\Omega Z\ . \end{array}$$

□

Either $\tilde{I}\circ\rho$ or $\gamma^{\natural\natural}$ may be used to write the homotopy equivalence

$$\Omega W_f \xrightarrow{\Delta} \Omega W_f \times \Omega W_f \xrightarrow{\Omega p\times\gamma^{\natural\natural}} \Omega Y \times \Omega^2 Z\ .$$

In the case where Y, Z are Eilenberg-Mac Lane spaces representing mod p cohomology, the Hopf algebra structure on

$$H^*(\Omega W_f; Z/p)$$

induced by loop multiplication may be worked out in terms of this splitting and the map induced by ζ of Prop. 3.5.4. In particular, if the maximal value of k is greater than 1, then classes coming from $\Omega^2 Z$ have the same coproduct as in their factor; there is no extension problem for the Hopf algebra structure.

Lest the point be lost in the details, ζ is a map we expect to compute, and Prop. 3.5.5 entails that the theory surrounding the Peterson-Stein formulas may be invoked.

Remark. The construction of γ used in Prop. 3.5.4 is an example of a box operation, introduced by McClendon [**76**]. The argument which expresses γ as a secondary composition could be promoted to a general proposition, but we do not pursue that line of thought here.

Chapter 4

Secondary Cohomology Operations

This chapter is devoted to the construction of secondary cohomology operations and their basic properties. We also treat operations of higher order. The principal novelty is in the passage from primary to secondary operations. The adaption of these ideas to higher orders amounts to modifications of the procedure used in the construction of secondary operations. Thus most of our discussion will concern secondary operations.

4.1. Definitions

To begin, we shall define secondary operations associated with a pair of composable maps

$$C_0 \xrightarrow{\theta} C_1 \xrightarrow{\varphi} C_2, \text{ with } C_2 \text{ a simply connected } H\text{-space.}$$

We first define the source and target.

Definition 4.1.1. Given a space X, let $S_\theta(X)$ denote the set of homotopy classes of maps $\varepsilon : X \to C_0$ such that the composition $\theta \circ \varepsilon$ is null-homotopic,

$$S_\theta(X) = \{[\varepsilon] \mid \varepsilon : X \to C_0,\ \theta \circ \varepsilon \sim *\}\ .$$

Continuing the definition, let $T_{\Omega\varphi}(X)$ denote the quotient,

$$T_{\Omega\varphi}(X) = [X, \Omega C_2]/im\Omega\varphi_\#$$

where

$$\Omega\varphi_\# : [X, \Omega C_1] \to [X, \Omega C_2]$$

is given by

$$\Omega\varphi_\#(g) = \Omega\varphi \circ g\ .$$

We write $\llbracket g \rrbracket$ to denote the image of $g : X \to \Omega C_2$ in $T_{\Omega\varphi}(X)$. We expect no confusion in this use of the letter T (a functor) with its use in denoting mapping cones.

For $f : Y \to X$, we have

$$f^{\#} : S_\theta(X) \to S_\theta(Y) \text{ given by}$$
$$f^{\#}[\varepsilon] = [\varepsilon \circ f]$$

and

$$f^{\#} : T_{\Omega\varphi}(X) \to T_{\Omega\varphi}(Y) \text{ given by}$$
$$f^{\#}\llbracket g \rrbracket = \llbracket g \circ f \rrbracket \,.$$

Definition 4.1.2. A secondary cohomology operation is a natural transformation of the functors

$$\Theta : S_\theta(\ \ ) \to T_{\Omega\varphi}(\ \ )\,.$$

For a space X, $S_\theta(X)$ is a set and $T_{\Omega\varphi}(X)$ is an abelian group. In the case where both θ and $\Omega\varphi$ are 0, our definition agrees with a primary cohomology operation.

If X is a point, Θ is the zero map. Hence $\Theta(\varepsilon) = 0$ for any $\varepsilon : X \to C_0$ which is null-homotopic. Thus Θ is automatically a map of pointed sets.

Next we discuss the semi-additivity property. Let (X, Y) be a pointed NDR pair. In subsection (1.3.3), we introduced a map

$$r : X/Y \to \Sigma Y\,.$$

Let

$$c : X/Y \to X/Y \vee \Sigma Y$$

be defined by the composition

$$X/Y \xrightarrow{r_1} X \cup T_1 Y \to X \cup T_1 Y \vee \Sigma Y \xrightarrow{P_0 \vee Id} X/Y \vee \Sigma Y$$

where the unnamed map in the middle is obtained by pinching

$$\left\{\frac{1}{2}\right\} \times Y \text{ in } T_1 Y$$

to a point, and p_0, r_1 are the inverse pair of homotopy equivalences from subsection (1.3.3).

Given maps

$$\varepsilon_1 : X/Y \to C_0,\ \varepsilon_2 : \Sigma Y \to C_0$$

we add them to form

$$\varepsilon_1 + \varepsilon_2 \circ r : X/Y \to C_0$$

by the composition

$$X/Y \xrightarrow{c} X/Y \vee \Sigma Y \xrightarrow{\varepsilon_1 \vee \epsilon_2} C_0 \vee C_0 \xrightarrow{\text{fold}} C_0 \,.$$

4.1.1. Semi-additivity. If ε_1 and ε_2 represent elements in $S_\theta(X/Y)$ and $S_\theta(\Sigma Y)$ respectively, then $\epsilon_1 + \epsilon_2 \circ r$ represents an element of $S_\theta(X/Y)$ and

$$\Theta(\varepsilon_1 + \varepsilon_2 \circ r) = \Theta(\varepsilon_1) + \Theta(\varepsilon_2 \circ r)$$

in $T_{\Omega\varphi}(X/Y)$.

Proof. We have the homeomorphism of pointed function spaces

$$\text{map}_*(X/Y \vee \Sigma Y, C_0) \cong \text{map}_*(X/Y, C_0) \times \text{map}_*(\Sigma Y, C_0) \,,$$

induced by the fold map $C_0 \vee C_0 \to C_0$.

It follows that

$$S_\theta(X/Y \vee \Sigma Y) = S_\theta(X/Y) \times S_\theta(\Sigma Y) \,.$$

Hence Θ is defined on $\varepsilon_1 + \varepsilon_2 \circ r$ by naturality. Furthermore, by naturality in the target variable, we have

$$T_{\Omega\varphi}(X/Y \vee \Sigma Y) = T_{\Omega\varphi}(X/Y) \times T_{\Omega\varphi}(\Sigma Y) \,.$$

Hence

$$\begin{aligned}\Theta(\varepsilon_1 + \varepsilon_2 \circ r) &= \Theta(\text{ fold } \circ (\varepsilon_1 \vee \varepsilon_2) \circ c) \\ &= c^{\#}(\Theta(\varepsilon_1), \Theta(\varepsilon_2)) \text{ in } T_{\Omega\varphi}(X/Y) \times T_{\Omega\varphi}(\Sigma Y) \\ &= \Theta(\varepsilon_1) + \Theta(\varepsilon_2 \circ r) \quad .\end{aligned}$$

□

4.1.2. Operations associated with null-homotopic compositions. To establish the existence of functions satisfying the definition (1.2) we make the assumption that there is a contracting homotopy H for the composition $\varphi \circ \theta$,

$$(\varphi, \theta, H) \text{ is a sequence with homotopy.}$$

Then with $\varepsilon : X \to C_0$ representing an element in $S_\theta(X)$, we have a set of secondary compositions

$$\{\tilde{\varphi} \circ \bar{\varepsilon} \mid \bar{\varepsilon} : X \to W_\theta \text{ is a lift of } \varepsilon\} \,.$$

From subheading (3.2.8), we obtain the information that the value of

$$[\![\tilde{\varphi} \circ \bar{\varepsilon}]\!] \text{ in } T_{\Omega\varphi}(X)$$

does not depend on the choice of lifting. From Prop. 3.3.7, it follows that these secondary compositions are invariants of the homotopy class (in the sense of Defn. 3.3.6) of the sequence with homotopy (φ, θ, H). Thus, there is a well-defined, natural transformation

$$\Theta : S_\theta(\quad) \to T_{\Omega\varphi}(\quad)$$

given by the formula

$$\Theta(\varepsilon) = [\![\tilde{\varphi} \circ \bar{\varepsilon}]\!] \ ,$$

for each homotopy class of (φ, θ, H) and we say Θ is <u>associated</u> to (φ, θ, H).

The diagram

$$\begin{array}{ccc} W_\theta & \xrightarrow{\tilde{\varphi}} & \Omega C_2 \\ \Big\downarrow{\scriptstyle p_0} & & \\ C_0 & \xrightarrow{\theta} & C_1 \end{array}$$

is known as the <u>universal example</u> for Θ associated with (φ, θ, H). This means that, in principle, the value of $\Theta(\varepsilon)$, for any $\varepsilon : X \to C_0$, is determined by

$$\Theta(p_\theta) \ .$$

Observe first that p_θ is in $S_\theta(W_\theta)$, and next that

$$\Theta(p_\theta) = [\![\tilde{\varphi}]\!] \ .$$

Thus

$$\Theta(\varepsilon) = \bar{\varepsilon}^{\#}\Theta(p_\theta)$$

where $\bar{\varepsilon}$ lifts ε.

The universal example is helpful to display the influence on the value of Θ by the choice of contracting homotopy H. From subheading (3.3.1) we have

$$\Theta_2(\varepsilon) = \Theta_1(\varepsilon) + \delta(H_1, H_2) \circ \varepsilon$$

for operations associated with (φ, θ, H_1) and (φ, θ, H_2). The equation holds in $T_{\Omega\varphi}(X)$.

In practice, we have maps (e.g. representing primary operations) φ, θ such that

$$[\varphi][\theta] = 0 \ .$$

We say a secondary operation Θ is <u>based on the relation</u> $[\varphi][\theta] = 0$ if Θ is associated to

$$(\varphi, \theta, H)$$

for some contracting homotopy H. We call H a <u>tethering</u> for Θ.

Now we work out these ideas for the example introduced in Chapter 2. Recall that we are concerned with distinguishing a space

$$S^n \cup e^{n+3}$$

from a bouquet of spheres. Here we illustrate the first part of that discussion — the construction of suitable operations.

We have, for a fixed integer n, the spaces

$$C_0 = K(Z/2, n) ,$$
$$C_1 = K(Z/2, n+1) \times K(Z/2, n+2) ,$$
$$C_2 = K(Z/2, n+4) ,$$
$$\theta : C_0 \to C_1 \text{ representing } \begin{pmatrix} Sq^1 \\ Sq^2 \end{pmatrix} ,$$
$$\varphi : C_1 \to C_2 \text{ representing } (Sq^3, Sq^2) ,$$

and

$$[\varphi][\theta] = 0 .$$

For a space X, $S_\theta(X)$ represents n-dimensional cohomology classes which are mapped to 0 by Sq^1 and Sq^2. Moreover, $T_{\Omega\varphi}(X)$ represents the quotient

$$H^{n+3}(X)/Sq^3 H^n + Sq^2 H^{n+1}$$

with $Z/2$ coefficients suppressed. Differences in tetherings are measured by maps

$$\delta : C_0 \to \Omega C_2 .$$

These maps are determined by cohomology classes in the subspaces

$$\text{span } \{Sq^3 b_n, Sq^2 Sq^1 b_n\} \qquad n \geq 3 ,$$
$$\text{span } \{Sq^2 Sq^1 b_2, b_2 \cup Sq^1 b_2\} \quad n = 2 ,$$
$$\text{span } \{Sq^2 Sq^1 b_1 = b_1^4\} \qquad n = 1 .$$

Since $Sq^3 = Sq^1 Sq^2$, these differences do not affect the value of Θ, because

$$\delta \circ \varepsilon \sim *$$

for any ε in $S_\theta(X)$.

The submodule used to form the quotient $T_{\Omega\varphi}(X)$ has an importance requiring a name. We call

$$im\, \Omega\varphi_\# , \ \Omega\varphi_\# : [X, \Omega C_1] \to [X, \Omega C_2]$$

the <u>indeterminacy</u> of Θ and write it as

$$Ind(\Theta, X) = im\, \Omega\varphi_\# .$$

Note that for $f : Y \to X$

$$f^\# Ind(\Theta, X) \subset Ind(\Theta, Y)$$

and it is the <u>latter</u> object that is used to form the quotient

$$T_{\Omega\varphi}(Y) = [Y, \Omega C_2]/Ind(\Theta, Y) .$$

In particular, indeterminancy is not governed by a universal example.

Remark. The semi-additivity property requires no use of tetherings. Furthermore, given a natural transformation Θ, we can choose a map $h : W_\theta \to \Omega C_2$ such that $\Theta(p_\theta) = [\![h]\!]$. This raises the question whether h is represented by a colifting. We defer these questions to Chapter 7. A partial answer is provided by item (iv) in the discussion of the structure theorems under subheading 7.2.1.

4.2. The fundamental theorem for secondary operations

We continue our discussion of the principal features of secondary operations under the assumption that the operation is based on a relation. Thus we are working with compositions

$$C_0 \xrightarrow{\theta} C_1 \xrightarrow{\varphi} C_2$$

where $[\varphi][\theta] = 0$, the space C_2 is a simply connected H-space, and each space has the homotopy type of a connected CW complex, and the initial spaces are either simply connected or H-spaces.

4.2.1. Loops and adjoints. The formula for adjoints

$$(\Omega\alpha \circ \varepsilon)^\flat = \alpha \circ \varepsilon^\flat$$

given the maps

$$X \xrightarrow{\varepsilon} \Omega H \xrightarrow{\Omega\alpha} \Omega B\ ,$$

was discussed in subsection (1.3.5) for primary operations, regarded simply as maps $\theta : C_0 \to C_1$. We use this approach to define the loop operation on a secondary operation. In these discussions involving the loop functor on spaces, we tacitly assume enough connectivity so as to be working with connected spaces after looping.

By taking adjoints, we have a set equivalence

$$S_{\Omega\theta}(X) = S_\theta(\Sigma X)$$

and an isomorphism of abelian groups,

$$T_{\Omega^2\varphi}(X) \cong T_{\Omega\varphi}(\Sigma X)\ .$$

Thus, given

$$\Theta : S_\theta(\ \ ) \to T_{\Omega\varphi}(\ \ )$$

we can define a natural transformation

$$\Omega\Theta : S_{\Omega\theta}(\ \ ) \to T_{\Omega^2\varphi}(\ \ )$$

by the diagram

$$\begin{array}{ccc} S_\theta(\Sigma X) & \xrightarrow{\Theta} & T_{\Omega\varphi}(\Sigma X) \\ \| & & \| \\ S_{\Omega\theta}(X) & \xrightarrow[\Omega\Theta]{} & T_{\Omega^2\varphi}(X) \, . \end{array}$$

In the notation of subheading (3.1.2), the definition of $\Omega\Theta$ reads,

$$\Omega\Theta(\varepsilon) = (\Theta(\varepsilon^\flat))^\natural \, .$$

The expression for this equation in terms of the Mayer-Vietoris suspension is

$$\Delta^* \Omega\Theta(y) = \Theta(\Delta^* y)$$

where y is represented by $\varepsilon : X \to \Omega C_0$.

Exercise 4.2.1. Let Θ be based on $[\varphi][\theta] = 0$. Prove

$$\begin{array}{ccc} S_\theta(X) & \xrightarrow{\Theta} & T_{\Omega\varphi}(X) \\ {\scriptstyle\sigma^*}\downarrow & & \downarrow{\scriptstyle\sigma^*} \\ S_{\Omega\theta}(\Omega X) & \xrightarrow[\Omega\Theta]{} & T_{\Omega^2\varphi}(\Omega X) \end{array}$$

commutes, where σ^* is the cohomology suspension.

In particular, it follows from the semi-additivity property that $\Omega\Theta$ is additive.

The phenomenon discussed in subheading (3.1.7) bears on universal examples and tetherings for $\Omega\Theta$. Suppose Θ is associated to

$$(\varphi, \theta, H) \, .$$

We have the sequence with homotopy

$$(\Omega\varphi, \ \Omega\theta, \ \Omega H)$$

but the associated secondary operation is not $\Omega\Theta$ in general. There is a sign from the switch of the variables map. We bring the discussion of subheading (3.2.7) to the present situation.

First, observe that

$$S_{T\circ\Omega\theta}(\qquad) = S_{\Omega\theta}(\qquad)$$

for the direction reversal map. Suppose

$$\begin{array}{ccc} W_\theta & \xrightarrow{\tilde{\varphi}} & \Omega C_2 \\ {\scriptstyle p_\theta}\downarrow & & \\ C_0 & \xrightarrow{\theta} & C_1 \end{array}$$

is the universal example for Θ associated to (φ, θ, H). Construct the universal example from the data

$$(\Omega\varphi,\ \tau \circ \Omega\theta,\ \tau \circ \Omega H)\ ,$$

$$\begin{array}{ccc} W_{\tau\circ\Omega\theta} & \xrightarrow{\widetilde{\Omega\varphi}} & \Omega^2 C_2 \\ {\scriptstyle p_{\tau\circ\Omega\theta}}\downarrow & & \\ \Omega C_0 & \xrightarrow[\tau\circ\Omega\theta]{} & \Omega C_1 \end{array} \quad .$$

From (3.2.7) we have a canonical fiber homotopy equivalence

$$\rho : \Omega W_\theta \to W_{\tau\circ\Omega\theta}$$

and

$$\widetilde{\Omega\varphi} \circ \rho \simeq \Omega\tilde{\varphi}\ .$$

Hence, we have

4.2.2. Sequence for $\Omega\Theta$. $\Omega\Theta$ is associated to

$$(\Omega\varphi,\ \tau \circ \Omega\theta,\ \tau \circ \Omega H)\ .$$

In particular, if ε is in $S_{\Omega\theta}(X)$, we have

$$\Omega\Theta(\varepsilon) = [\![\Omega\tilde{\varphi} \circ \varepsilon_1]\!] = [\![\widetilde{\Omega\varphi} \circ \varepsilon_2]\!]$$

in $T_{\Omega^2\varphi}(X)$, where

$$\Omega p_\theta \circ \varepsilon_1 \simeq \varepsilon \simeq p_{\tau\circ\Omega\theta} \circ \varepsilon_2\ .$$

4.2.3. Compatibility with exact sequences (cf. [21] exposé 13). Let $f : X \to Y$ and consider the sequence

$$(*) \qquad X \xrightarrow{f} Y \xrightarrow{j} T_f \xrightarrow{q} \Sigma X\ ,$$

where j is the inclusion and q is obtained by pinching Y to a point in T_f. Let Θ be an operation based on the relation $[\varphi][\theta] = 0$, where, in addition, we assume that φ is an H-map. This section provides means for gaining information on the values of

$$\Theta : S_\theta(Y) \to T_{\Omega\varphi}(Y)$$

in terms of φ, θ, and the sequence $(*)$. The result is independent of tetherings.

First we define the maps

$$A : [T_f, C_0] \longrightarrow [Y, C_1]$$

given by

$$A(\varepsilon) = \theta \circ \varepsilon \circ j ,$$

and

$$B : [\Sigma Y, C_1] \longrightarrow [\Sigma X, C_2]$$

given by

$$B(\gamma) = \varphi \circ \gamma \circ \Sigma f .$$

Then A is a map of pointed sets and B is a homomorphism of abelian groups We have the display

$$\begin{array}{ccccc}
 & & [\Sigma Y, C_1] & = & [\Sigma Y, C_1] \\
 & & {\scriptstyle (\Sigma f)^{\#}}\downarrow & & \downarrow{\scriptstyle B} \\
 & & [\Sigma X, C_1] & \xrightarrow[\varphi_{\#}]{} & [\Sigma X, C_2] \\
 & & {\scriptstyle q^{\#}}\downarrow & & \\
[T_f, C_0] & \xrightarrow{\theta_{\#}} & [T_f, C_1] & & \\
\| & & {\scriptstyle j^{\#}}\downarrow & & \\
[T_f, C_0] & \xrightarrow[A]{} & [Y, C_1] . & &
\end{array}$$

From the exactness properties of the middle column, we obtain a well-defined map, depending only on φ and θ,

$$\Delta = \Delta(\varphi, \theta) : \ker A \to \operatorname{coker} B$$

given by

$$\Delta = \varphi_{\#} \circ (q^{\#})^{-1} \circ \theta_{\#} .$$

Now $j^{\#}$ maps ker A to $S_\theta(Y)$ and $(\Sigma f)^{\#}$ maps $T_{\Omega\varphi}(Y) = T_\varphi(\Sigma Y)$ to coker B. The property we call "compatibility with exact sequences" is given by

Proposition 4.2.2. The following diagram commutes,

$$\begin{array}{ccc}
\ker A & \xrightarrow{\Delta} & \operatorname{coker} B \\
{\scriptstyle j^{\#}}\downarrow & & \uparrow{\scriptstyle -(\Sigma f)^{\#}} \\
S_\theta(Y) & \xrightarrow[\Theta]{} & T_{\Omega\varphi}(Y) = T_\varphi(\Sigma Y) ,
\end{array}$$

where the direction reversal map is indicated by a minus sign.

Proof. We use the universal example for Θ and the basic Peterson-Stein formula (3.4.2) in an argument like that of Prop. 3.5.2.

Let

$$\varepsilon : T_f \to C_0$$

map to an element in $\ker A$. Choose

$$\zeta : \Sigma X \to C_1$$

such that

$$\theta \circ \varepsilon \simeq \zeta \circ q \, .$$

The coextension

$$\underset{\sim}{f} : \Sigma X \to T_j$$

is a homotopy equivalence, so there is a map

$$\zeta' : T_j \to C_1$$

such that

$$\zeta \simeq (-1) \text{ times } \zeta' \circ \underset{\sim}{f} \, .$$

Thus we can form the following diagram, with homotopy commutative squares, except on the left in the top row, which anti-commutes, indicated by (-1),

$$\begin{array}{ccccccc}
 & & T_f & \xrightarrow{q} & \Sigma X & \xrightarrow{-\Sigma f} & \Sigma Y \\
 & & \| & \boxed{-1} & \downarrow{\scriptstyle \underset{\sim}{f}} & & \downarrow{\scriptstyle \underset{\sim}{j}} \\
Y & \xrightarrow{j} & T_f & \xrightarrow{\hat{j}} & T_j & \xrightarrow{h} & T_{\hat{j}} \\
\downarrow{\scriptstyle \overline{\varepsilon j}} & & \downarrow{\scriptstyle \varepsilon} & & \downarrow{\scriptstyle \zeta'} & & \downarrow{\scriptstyle \underline{\varphi\zeta'}} \\
W_\theta & \longrightarrow & C_0 & \xrightarrow[\theta]{} & C_1 & \xrightarrow[\varphi]{} & C_2 \\
\downarrow{\scriptstyle \tilde{\varphi}} & & & & & & \\
\Omega C_2 \, . & & & & & &
\end{array}$$

The outer maps $\overline{\varepsilon j}$ and $\underline{\varphi\zeta'}$ are constructed by following the instructions in subheading (3.4.1).

By definition

$$\Delta(\varepsilon) = [\![\varphi \circ \zeta]\!] \text{ in coker } B \, .$$

On the other hand,

$$-(\Sigma f)^{\#}\Theta(\varepsilon \circ j)^{\flat} = -[\![(\tilde{\varphi} \circ \overline{\varepsilon j})^{\flat} \circ \Sigma f]\!] \, .$$

By the basic Peterson-Stein formula (3.4.2), the term on the right is the image in coker B of

$$\underline{\varphi\zeta'} \circ \underset{\sim}{j} \circ \Sigma f \simeq -\varphi \circ \zeta' \circ \underset{\sim}{f} \simeq \varphi \circ \zeta \,.$$

□

When we invoke Prop. 4.2.2, we use the phrase "Θ is compatible with Δ." In particular, $\Omega\Theta$ is compatible with $\Delta(\Omega\varphi, \tau \circ \Omega\theta)$, and

$$\Delta(\Omega\varphi\,,\ \tau \circ \Omega\theta) = -\Delta(\Omega\varphi,\ \Omega\theta)\,.$$

We restate Prop. 4.2.2 in terms of classical cohomology for an NDR pair (Y, X) with f the inclusion of the subspace and j the inclusion of Y in the pair. Suppose ε, ζ are cohomology classes respectively for (Y, X) and X and Θ is defined on $j^*\varepsilon$. Suppose $\theta(\varepsilon) = \delta(\zeta)$ where δ is the connecting homomorphism. Then

$$f^*\Theta(j^*\varepsilon) = [\![\Omega\varphi(\zeta)]\!] \text{ taken modulo } f^*Ind(\Theta, Y)\,.$$

To prove this, superimpose $r : Y/X \to \Sigma X$ over q in the proof of Prop. 4.2.2. We obtain the stated equation using Propositions 1.3.1, 1.3.2, and the remark in subsection (3.2.7). Essentially r and q differ by a sign. □

We return to the discussion of Chapter 2, to use compatibility with exact sequences to make a calculation. To set the stage, we have

$$\eta_n : S^{n+1} \to S^n$$

which is detected by Sq^2. We ask (again) whether the composition

$$\eta_n \circ \eta_{n+1} = \eta^2 : S^{n+2} \to S^n$$

is essential. The new approach is through the mapping cone, which we wish to distinguish from a bouquet of spheres. Let $Y = T_{\eta^2}$. To employ Prop. 4.2.2 we need a suitable exact sequence. The following construction is a special case of a general construction for compositions.

The diagram below defines spaces and maps

$$X = T_{\eta_{n+1}},\ Y = T_{\eta^2} \text{ and } f : X \to Y\,,$$

$$\begin{array}{ccccc}
S^{n+2} & = & S^{n+2} & & \\
\downarrow{\scriptstyle \eta_{n+1}} & & \downarrow{\scriptstyle \eta^2} & & \\
S^{n+1} & \xrightarrow{\eta_n} & S^n & \longrightarrow & T_{\eta_n} \\
\downarrow & & \downarrow & & \downarrow \\
T_{\eta_{n+1}} & \xrightarrow[f]{} & T_{\eta^2} & \xrightarrow[j]{} & T_f\ ,
\end{array}$$

and reveals a homotopy equivalence of T_{η_n} with T_f. Recall that Θ is based on $\varphi\theta = 0$ where

$$\theta = \begin{pmatrix} Sq^1 \\ Sq^2 \end{pmatrix}\ , \ \varphi = (Sq^3\ , \ Sq^2)$$

and

$$S_\theta(Y) = H^n(T_{\eta^2})\ ,$$

$$T_{\Omega\varphi}(Y) = H^{n+3}(T_{\eta^2}),\ (Z/2 \text{ coefficients})\ .$$

We have

$$A : H^n(T_f) \to H^{n+1}(Y) \oplus H^{n+2}(Y) \text{ is the zero map,}$$

and

$$B : H^{n+1}(\Sigma Y) \oplus H^{n+3}(\Sigma Y) \to H^{n+4}(\Sigma X) \text{ is the zero map.}$$

Furthermore,

$$Sq^2 : H^n(T_f) \to H^{n+2}(T_f)$$

is an isomorphism, because Sq^2 detects η and there is identification of T_f with T_η. Moreover,

$$q^\# : H^{n+2}(\Sigma X) \to H^{n+2}(T_f)$$

is an isomorphism for dimensional reasons, and

$$Sq^2 : H^{n+2}(\Sigma X) \to H^{n+4}(\Sigma X)$$

is an isomorphism. Hence $\Delta = \Delta(\varphi, \theta) \neq 0$, and likewise for Θ.

On the other hand, if there were a homotopy equivalence

$$T_f \simeq S^n \vee S^{n+3}\ ,$$

then we have $\varepsilon : T_f \to K(Z/2, n)$ given by

$$T_f \xrightarrow{g} S^n \xrightarrow{\varepsilon'} K(Z/2, n)$$

where g is the composition of a putative equivalence and a retraction, and ε is essential. But

$$\Theta(\varepsilon') = 0 \text{ for dimensional reasons}$$

which contradicts the above calculation of Θ. Hence η^2 is essential.

Remark. The argument distinguishing T_f from $S^n \vee S^{n+3}$ cannot be based solely on the construction of essential secondary compositions. The indeterminancy is also required. The point is illustrated by example.

Let

$$W = (S^n \vee S^{n+1}) \bigcup_{(\eta^2, \eta)} e^{n+3} ,$$

with the top cell attached non-trivially to both spheres. We have a map

$$g : W \to X$$

obtained by pinching S^{n+1} to a point. Moreover, the induced map $g^* : H^{n+3}(X) \to H^{n+3}(W)$ is an isomorphism. Let

$$\varepsilon : X \to K(Z/2, n)$$

represent a generator, then, for any secondary composition

$$\tilde{\varphi} \circ \bar{\varepsilon}$$

representing $\Theta(\varepsilon)$, we have the fact that

$$\tilde{\varphi} \circ \bar{\varepsilon} \circ g \text{ is essential.}$$

Of course

$$[\![\tilde{\varphi} \circ \bar{\varepsilon} \circ g]\!] = 0 \text{ in } T_{\Omega\varphi}(W) ,$$

so no conclusion may be drawn whether S^n splits off W.

Remark. The fact that η^2 has order 2 may be used to construct a self-equivalence

$$\begin{pmatrix} 1 & 0 \\ \eta, & 1 \end{pmatrix} : S^n \vee S^{n+1} \to S^n \vee S^{n+1} .$$

By use of this equivalence to reattach the top cell, we obtain

$$W \simeq S^n \vee Y .$$

Remark. We have distinguished T_{η^2} from a bouquet of spheres but, so far, have not ruled out other non-trivial 2-cell complexes of the form $S^n \cup e^{n-3}$. In fact $\pi_{n+2}S^n = Z/2$ for $n \geq 2$, so there are no other homotopy types.

Remark. In Madsen and Milgram [**62**], p. 32 the reader can find interesting uses of the fact that η^2 is detected by the secondary operations worked out above. Another, recent, paper using this information is Kitchloo and Shankar [**53**].

4.2.4. Omnibus Theorem. We collect the main points of the two previous sections under

Fundamental Theorem for Secondary Operations. *Associated to each sequence with homotopy*

$$(\varphi, \theta, H),\ \theta : C_0 \to C_1, \varphi : C_1 \to C_2$$

there is a natural transformation

$$\Theta : S_\theta(\quad) \to T_{\Omega\varphi}(\quad)$$

satisfying:

- **(a):** *homotopy invariance:* *If* (φ, θ, H) *is homotopic to* (φ', θ', H'), *then the associated natural transformations* Θ *and* Θ' *are equal.*
- **(b):** *dependence on the homotopy:* *If* Θ, Θ' *are associated to* (φ, θ, H) *and* (φ', θ', H') *respectively and* $[\varphi] = [\varphi']$, $[\theta] = [\theta']$, *then for any* ε *in* $S_\theta(X)$,

$$\Theta(\varepsilon) - \Theta'(\varepsilon) = [\![\delta \circ \varepsilon]\!]$$

for some $\delta : C_0 \to \Omega C_2$. *If* $\varphi = \varphi'$ *and* $\theta = \theta'$, *then* δ *can be chosen as*

$$\delta = \delta(H, H')\ .$$

- **(c):** *semi-additivity:* *Let* (X, Y) *have the homotopy type of a CW pair,* u *and* v *be elements in* $S_\theta(X/Y)$ *and* $S_\theta(\Sigma Y)$ *respectively. Then* $u + v \circ r$ *is an element of* $S_\theta(X/Y)$ *and*

$$\Theta(u + v \circ r) = \Theta(u) + \Theta(v \circ r)$$

where $r : X/Y \to \Sigma Y$. *In particular,* Θ *is additive on* $S_\theta(\Sigma X)$.

- **(d):** *looping:* *If* $\Omega\Theta$ *is obtained from* Θ *by adjoints,*

$$\Omega\Theta(\epsilon) = (\Theta(\varepsilon^\flat))^\natural\ ,$$

then $\Omega\Theta$ *is associated to* $(\Omega\varphi, \tau\circ\Omega\theta, \tau\circ\Omega H)$ *where* τ *is the direction reversal map.*

- **(e):** *compatibility with exact sequences:* *If* $f : X \to Y$ *and* φ *is an* H*-map, then the following diagram commutes:*

$$\begin{array}{ccc} \ker A & \xrightarrow{\ \Delta\ } & coker\, B \\ {\scriptstyle j^\#}\downarrow & & \uparrow{\scriptstyle -(\Sigma f)^\#} \\ S_\theta(Y) & \xrightarrow{\ \Theta\ } & T_{\Omega\varphi}(Y) = T_\varphi(\Sigma Y)\ , \end{array}$$

where Δ *is defined in terms of* θ *and* φ.

(f): *universal example:* *On elements ε in $S_\theta(X)$, the value of Θ is given by*

$$\Theta(\varepsilon) = [\![\tilde{\varphi} \circ \bar{\varepsilon}]\!]$$

where $\tilde{\varphi}$ is the colifting constructed from (φ, θ, H) and $\bar{\varepsilon}$ is any lifting of ε. Furthermore, on elements ε in $S_{\Omega\theta}(X)$, $\Omega\Theta$ is given by

$$\Omega\Theta(\varepsilon) = [\![\Omega\tilde{\varphi} \circ \varepsilon_1]\!] = [\![\widetilde{\Omega\varphi} \circ \varepsilon_2]\!]$$

$\varepsilon_1 : X \to \Omega W_\theta$, $\varepsilon_2 : X \to W_{\tau\circ\Omega\theta}$ are lifts of $\varepsilon : X \to \Omega C_0$ and $\widetilde{\Omega\varphi}$ is the colifting constructed from $(\Omega\varphi, \tau \circ \Omega\theta, \tau \circ \Omega H)$.

The remainder of this subsection and the next subsection are offered as four extended exercises.

Exercise 4.2.3. The pointed function space $\text{map}_*(X, Y)$ is abbreviated to $\langle X, Y\rangle$ in the following statements, and $\langle X, Y\rangle_f$ is the path component of a map $f : X \to Y$. Consider the following pull-back fiber squares,

$$\begin{array}{ccccc}
W_{(\varphi,\theta)} & \longrightarrow & W & \longrightarrow & \langle C_0, P_0C_2\rangle \\
\downarrow & & \downarrow & & \downarrow{\scriptstyle e_1} \\
\langle C_1, C_2\rangle_\varphi \times \langle C_0, C_1\rangle_\theta & \longrightarrow & \langle C_1, C_2\rangle \times \langle C_0, C_1\rangle & \longrightarrow & \langle C_0, C_2\rangle
\end{array}$$

where e_1 is evaluations at $\{1\}$ and the maps on the lower line are compositions of functions. Verify the following:

(a): $W_{(\varphi,\theta)}$ is the set of sequences with homotopy (φ, θ, H).

(b): (φ, θ, H) is homotopic to (φ', θ', H') if and only if they are in the same path component of $W_{(\varphi,\theta)}$.

(c): Let $G = \pi_0\langle C_0, \Omega C_2\rangle$, then G acts transitively on the set $\pi_0 W_{(\varphi,\theta)}$.

Exercise 4.2.4 (Functional Cohomology Operations)**.** These operations may be constructed with the methods of this chapter. Some of the basic points are developed here. Historically, the Peterson-Stein formulas were discovered in this context.

Consider a sequence

$$X \xrightarrow{f} Y \xrightarrow{\varepsilon} C_0 \xrightarrow{\theta} C_1$$

such that $\varepsilon \circ f \sim *$, $\theta \circ \varepsilon \sim *$ and C_1 is a simply-connected H-space, (but we do not assume θ is additive). By definition, the functional cohomology operation $\theta_f(\varepsilon)$ is given by

$$\theta_f(\varepsilon) = [\![\gamma]\!] \text{ in } [X, \Omega C_1]/imf^{\#} + im\Omega\theta_{\#} \;,$$

where $\gamma : X \to \Omega C_1$ satisfies $j \circ \gamma \simeq \bar{\varepsilon} \circ f$ and $\bar{\varepsilon} : Y \to W_\theta$ lifts ε. The diagramatic display is

$$
\begin{array}{ccccccc}
 & & \Omega C_1 & \xrightarrow{j} & W_\theta & & \\
 & \overset{\gamma}{\nearrow} & & \overset{\bar{\varepsilon}}{\nearrow} & \downarrow & & \\
X & \xrightarrow[f]{} & Y & \xrightarrow{\varepsilon} & C_0 & \xrightarrow[\theta]{} & C_1 \, .
\end{array}
$$

Without appealing to the H-structure on C_1, we have the following fact. If $j \circ \gamma_1 \simeq \varepsilon_1 \circ f$ and $j \circ \gamma_2 \simeq \varepsilon_2 \circ f$ with each ε_i lifting ε, then there are maps $\delta_1 : X \to \Omega C_0$ and $\delta_2 : Y \to \Omega C_1$ such that

$$\Omega\theta \circ \delta_1 + \delta_2 \circ f + \gamma_1 = \gamma_2 \text{ in } [X, \Omega C_1] \, .$$

Let $\Theta : S_\varepsilon(\quad) \to T_{\Omega\theta}(\quad)$ be a secondary operation based on $[\theta][\varepsilon] = 0$. Show

(a) $$\theta_f(\varepsilon) = -[\![\Theta(f)]\!] \text{ in } T_{\Omega\theta}(X)/imf^{\#} \, .$$

If $g : Z \to X$, show

$$\theta_{f \circ g}(\varepsilon) = g^{\#}\theta_f(\varepsilon) \text{ in } [Z, \Omega C_1]/im\Omega\theta_{\#} + im(f \circ g)^{\#} \, .$$

Suppose $f : X \to Y$ and $\theta : C_0 \to C_1$ are given. Set

$$Dom(Y) = \{[\varepsilon] \mid \varepsilon : Y \to C_0 \, , \; \varepsilon \circ f \sim *, \; \theta \circ \varepsilon \sim *\} \, .$$

Show that if θ is an H-map of H-spaces, then for any $\varepsilon_1, \varepsilon_2$ in $Dom(Y)$

(b) $$\theta_f(\varepsilon_1 + \varepsilon_2) = \theta_f(\varepsilon_1) + \theta_f(\varepsilon_2) \text{ in } [X, \Omega C_1]/imf^{\#} + im\Omega\theta_{\#} \, .$$

In analogy with Prop. 4.2.2, define the following functions,

$$A_f : [T_f, C_0] \to [Y, C_1], \qquad \text{a map of pointed sets given by}$$
$$A_f(\varepsilon) = \theta\varepsilon j$$

and likewise $B_f : [\Sigma X, C_0] \to [T_f, C_1]$, given by

$$B_f(\zeta) = \theta \circ \zeta \circ q \, .$$

The cooperation $c : T_f \to \Sigma X \vee T_f$ induces an action

$$[\Sigma X, C_0] \times [T_f, C_1] \to [T_f, C_1]$$

through θ. We can use this action to define an equivalence relation $\mathcal{R}$ on $[T_f, C_1]$. Show that the following diagram commutes,
(c)

$$\begin{array}{ccccc}
\ker A_f & \xrightarrow{\theta_\#} & [T_f, C_1] & \longrightarrow & [T_f, C_1]/\mathcal{R} \\
{\scriptstyle j^\#}\downarrow & & & & \uparrow{\scriptstyle (q\circ\tau)^\#} \\
Dom(Y) & \xrightarrow[\theta_f(\ )]{} & [X, \Omega C_1]/imf^\# + im\Omega\theta_\# & = & [\Sigma X, C_1]/im\Sigma f^\# + im\theta_\#
\end{array}$$

and $(q \circ \tau)^\#$ is an injection. We say that θ detects f if the top row is non-trivial. If the image of θ is "central" in C_1, i.e.

$$\begin{array}{ccccc}
C_0 \times C_1 & \xrightarrow{\theta\times 1} & C_1 \times C_1 & \xrightarrow{\text{mult.}} & C_1 \\
\downarrow & & & & \| \\
C_1 \times C_0 & \xrightarrow[1\times\theta]{} & C_1 \times C_1 & \xrightarrow[\text{mult.}]{} & C_1
\end{array}$$

is homotopy commutative, then

$$[T_f, C_1]/\mathcal{R} = \text{ coker } B_f \ .$$

(d): First Peterson-Stein formula. Suppose $\varphi : C_1 \to C_2$ satisfies $\varphi \circ \theta$ is null-homotopic. Let Θ denote any secondary operation based on $[\varphi][\theta] = 0$. Prove

$$f^\#\Theta(\varepsilon) = \Omega\varphi_\#(\theta_f(\varepsilon)) \text{ in } [X, \Omega C_2]/f^\# Ind(\Theta, Y) \ .$$

This formula finds applications in the evaluations of secondary operations in total spaces of fibrations. The diagram below displays a typical situation:

$$\begin{array}{ccccccc}
 & & E & \xrightarrow{j} & E/F & \longrightarrow & \Sigma F \\
 & & \| & & \downarrow{\scriptstyle \underline{p}} & & \downarrow \\
F & \xrightarrow{i} & E & \xrightarrow{p} & B & & {\scriptstyle V} \\
 & & \downarrow & & \downarrow{\scriptstyle \varepsilon} & & \downarrow \\
 & & W_\theta & \longrightarrow & C_0 & \xrightarrow[\theta]{} & C_1 \\
 & & {\scriptstyle \tilde{\varphi}}\downarrow & & & & \\
 & & \Omega C_2 \ . & & & &
\end{array}$$

One has in mind that $\varepsilon \circ p$ represents a class in the cohomology of E coming from the base B and one knows the action of θ in terms of the square on the right. Then $\theta_i(\varepsilon p)$ is known and is (-1) times V. Thus

$$i^{\#}\Theta(\varepsilon p) = -[\![\varphi \circ V^{\natural}]\!] \text{ in } [F, \Omega C_2]/i^{\#}Ind(\Theta, E) .$$

Now we may have other classes $\psi : E \to C_0$ with the same restriction to the fiber. In that case, the exact sequence for a fibration can be employed to give information on the relation of ψ with $\Theta(\varepsilon p)$.

Here is an example. It arises in certain calculations leading to the structure of $\pi_{n+3}S^n$, and $n \geq 5$. We write Sq^i as $Sq(i)$ and likewise for compositions:

$$\begin{array}{ccccccc}
K(n+1) & \longrightarrow & W_{Sq(2)} & \longrightarrow & W/K(n+1) & \longrightarrow & \Sigma K(n+1) \\
\| & & \| & & \downarrow & & \downarrow \\
F & \longrightarrow & W_{Sq(2)} & \xrightarrow{p} & K(Z,n) & \xrightarrow{Sq(2)} & K(n+2) \\
 & & \downarrow & & \downarrow{\scriptstyle Sq(4)} & & \downarrow{\scriptstyle Sq(2,1)} \\
 & & W_{Sq(1)} & \longrightarrow & K(n+4) & \xrightarrow[Sq(1)]{} & K(n+5) \\
 & & {\scriptstyle \widetilde{Sq(1)}}\downarrow & & & & \\
 & & K(n+5) \ . & & & &
\end{array}$$

Let $\Phi : S_{Sq(2)}(\ ) \to T_{Sq(2)}(\ )$ be a secondary operation based on $Sq(2)Sq(2) = 0$ (integrally). Then, using the Adem relation

$$Sq(1)Sq(2,1) = Sq(3,1) = Sq(2)Sq(2) ,$$

we obtain

$$i^{\#}Sq(2)\Phi(p) = i^{\#}\Theta(Sq(4)p) \text{ in}$$

$$[F, K(n+5)] \Big/ i^{\#}Ind(\Theta, W) + i^{\#}Sq(2)Ind(\Phi, W) .$$

(e): Second Peterson-Stein formula. Suppose we have a sequence

$$X \xrightarrow{f} Y \xrightarrow{\varepsilon} C_0 \xrightarrow{\theta} C_1 \xrightarrow{\varphi} C_2$$

with C_2 a simply connected H-space and $[\varphi][\theta] = 0$. Suppose

$$\varepsilon \circ f \text{ is in } S_\theta(X) .$$

Prove

$$\Theta(\varepsilon \circ f) = -\varphi_f(\theta \circ \varepsilon) \text{ in } [X, \Omega C_2] \Big/ im f^{\#} + im \Omega\theta_{\#} ,$$

where Θ is a secondary operation based on $[\varphi][\theta]=0$. This formula is the original form which evolved to the formulation in Chapter 3, section 4.

We have a formal statement, in the form of a diagram (c), when θ detects a map f. Here we develop an analogous diagram for secondary operations. Consider a sequence

$$X \xrightarrow{f} Y \xrightarrow{\varepsilon} C_0 \xrightarrow{\theta} C_1 \xrightarrow{\varphi} C_2$$

where each composition is null, C_1 and C_2 are H-spaces but φ need not be additive. We have a map

$$A_f : [T_f, W_\theta] \to [Y, \Omega C_2]$$

given by

$$A(\gamma) = \tilde{\varphi} \circ \gamma \circ j$$

where we have fixed a colifting $\tilde{\varphi}$.

If $[\Sigma X, C_0] = 0$, then the equivalence relation $\mathcal{R}$ on $[T_f, \Omega C_2]$ induced by the cooperation is passage to the quotient

$$[T_f, \Omega C_2] \Big/ im(\Omega\varphi_\# q^\#)\ .$$

Write

$$Dom(Y) = \{[\varepsilon] | \varepsilon : Y \to W_\theta,\ \varepsilon \circ f \sim *,\ \tilde{\varphi} \circ \varepsilon \sim *\}\ ,$$

$$co\ Dom(X) = [X, \Omega^2 C_2] \Big/ im f^\# + im \Omega\tilde{\varphi}_\#\ .$$

Then we have the following commutative diagram:

$$\text{(f)}\qquad \begin{array}{ccccc}
S_\theta(T_f) & & \xrightarrow{\quad\Theta\quad} & & T_{\Omega\varphi}(T_f) \\
\uparrow & & & & \uparrow{\scriptstyle proj.} \\
\ker A_f & \xrightarrow{\tilde{\varphi}_\#} & [T_f, \Omega C_2] & \longrightarrow & [T_f, \Omega C_2]/im(\Omega\varphi_\# q^\#) \\
{\scriptstyle j^\#}\downarrow & & & & \uparrow{\scriptstyle (q\circ\tau)^\#} \\
Dom(Y) & \xrightarrow[\tilde{\varphi}_f(\ \ )]{} & co\ Dom(X) & = & co\ Dom(\Sigma X)\ .
\end{array}$$

If $j^\# : [T_f, \Omega C_1] \to [Y, \Omega C_1]$ is the zero map, then the map marked "proj." is an isomorphism.

Both the Adem argument and compatibility with exact sequences provide means for deciding whether a composition is essential. Of course, the details depend on choosing elements, but there is a formal side to the arguments which can be codified in diagrams.

The Adem argument for a composition $X \xrightarrow{f} Y \xrightarrow{g} Z$ works from the diagram

$$\begin{array}{ccccc}
 & & \Sigma X & = & \Sigma X \\
 & & \downarrow{\scriptstyle \underset{\sim}{f}} & & \downarrow{\scriptstyle \Sigma f} \\
Z & \xrightarrow{j_1} & T_g & \xrightarrow{q_1} & \Sigma Y \\
\| & & \downarrow{\scriptstyle j_2} & & \downarrow{\scriptstyle j_3} \\
Z & \longrightarrow & T_{\underset{\sim}{f}} & \xrightarrow{q_4} & T_{\Sigma f} \\
 & & \downarrow{\scriptstyle q_2} & & \downarrow{\scriptstyle q_3} \\
 & & \Sigma^2 X & = & \Sigma^2 X
\end{array}$$

which can be constructed on the assumption that $g \circ f$ is null-homotopic. We introduce the following maps:

$$\begin{aligned}
A_g : [T_g, C_0] &\longrightarrow [Z, C_1]\ ,\ A_g(\gamma) = \theta \circ \gamma \circ j_1\ ,\\
B_g : [\Sigma Y, C_0] &\longrightarrow [T_g, C_1]\ ,\ B_g(\gamma) = \theta \circ \gamma \circ q_1\ ,\\
A_{\Sigma f} : [T_{\Sigma f}, C_1] &\longrightarrow [\Sigma Y, C_2]\ ,\ T_{\Sigma f}(\gamma) = \varphi \circ \gamma \circ j_3\ ,\\
B_{\Sigma f} : [\Sigma^2 X, C_1] &\longrightarrow [T_{\Sigma f}, C_2]\ ,\ B_{\Sigma f}(\gamma) = \varphi \circ \gamma \circ q_3\ ,\\
A_{\underline{g}} : [T_{\underset{\sim}{f}}, C_0] &\longrightarrow [Z, C_1]\ ,\ A_{\underline{g}}(\gamma) = \theta \circ \gamma \circ j_2 \circ j_1\ ,\\
B_{\underset{\sim}{f}} : [\Sigma^2 X, C_1] &\longrightarrow [T_{\underset{\sim}{f}}, C_2]\ ,\ B_{\underset{\sim}{f}}(\gamma) = \varphi \circ \gamma \circ q_2\ .
\end{aligned}$$

Then we have the following diagram, where the bottom middle square anti-commutes.

(g)

$$\begin{array}{ccccccc}
\ker A_{\underline{g}} & \xrightarrow{\theta_\#} & [T_{\underset{\sim}{f}}, C_1] & = & [T_{\underset{\sim}{f}}, C_1] & \xrightarrow{\varphi_\#} & \operatorname{coker} B_{\underset{\sim}{f}} \\
\downarrow{\scriptstyle j_2^\#} & & \downarrow{\scriptstyle j_2^\#} & & \uparrow{\scriptstyle q_4^\#} & & \uparrow{\scriptstyle q_4^\#} \\
\ker A_g & \xrightarrow{\theta_\#} & \operatorname{coker} B_g & \xleftarrow[\boxed{-1}]{} & \ker A_{\Sigma f} & \xrightarrow{\varphi_\#} & \operatorname{coker} B_{\Sigma f} \\
\downarrow{\scriptstyle j_1^\#} & & \uparrow{\scriptstyle (q_1 \circ \tau)^\#} & & \downarrow{\scriptstyle j_3^\#} & & \uparrow{\scriptstyle (q_3 \circ \tau)^\#} \\
Dom(Z) & \xrightarrow[\theta_g(\ \ )^\flat]{} & co\,Dom(\Sigma Y) & \longleftarrow & Dom(\Sigma Y) & \xrightarrow[\varphi_{\Sigma f}(\ \ )^\flat]{} & co\,Dom(\Sigma^2 X)
\end{array}$$

In practice, this set-up comes with the information that the intersection of the images of $\theta_\#$ and $q_4^\#$ in the top row is not empty.

Compatibility with exact sequences works from the diagram:

$$\begin{array}{ccccc}
X & = & X & & \\
{\scriptstyle f}\downarrow & & \downarrow & & \\
Y & \xrightarrow{g} & Z & \xrightarrow{j} & T_g \\
\downarrow & & \downarrow & & \downarrow \\
T_f & \xrightarrow{\hat{g}} & T_{gf} & \xrightarrow{\hat{j}} & T_{\hat{g}}
\end{array}$$

and the observation that the map from T_g to $T_{\hat{g}}$ is a homotopy equivalence. Here $\hat{g}$ is the extension of g induced by f. Then we have the following diagram:
(h)

$$\begin{array}{ccccccc}
S_\theta(T_{gf}) & & \xrightarrow{\Theta} & & T_{\Omega\varphi}(T_{gf}) & = & T_\varphi(\Sigma T_{gf}) \\
{\scriptstyle \hat{j}^\#}\uparrow & & & & & & \downarrow{\scriptstyle -(\Sigma\hat{g})^\#} \\
\ker A_{\hat{g}} & \xrightarrow{\theta_\#} & [T_{\hat{g}}, C_1] & \xleftarrow{\hat{q}^\#} & [\Sigma T_f, C_1] & \longrightarrow & \text{coker } B_{\hat{g}} \\
\downarrow & & \downarrow & & \uparrow & & \uparrow \\
\ker A_g & \xrightarrow{\theta_\#} & \text{coker } B_g & \xleftarrow[\boxed{-1}]{} & \ker A_{\Sigma f} & \xrightarrow{\varphi_\#} & \text{coker } B_{\Sigma f} \\
\downarrow & & \uparrow & & \downarrow & & \uparrow \\
Dom(Z) & \xrightarrow[\theta_g(\ \ )^\flat]{} & co\ Dom(\Sigma Y) & \longleftarrow & dom(\Sigma Y) & \xrightarrow[\varphi_{\Sigma f}(\ \ )^\flat]{} & co\ Dom(\Sigma^2 X)
\end{array}$$

Note that $\varphi_\# \circ (\hat{q}^\#)^{-1} \circ \theta_\# = \Delta(\varphi, \theta)$.

In practice, this set-up comes with additional information, so that $\ker A_{\Sigma f}$ maps to suitable elements of $(\hat{q}^\#)^{-1}$ on $im\ \theta_\#$.

Both diagrams (g) and (h) reveal composition formulas for functional operations. These matters and more are treated in [**89**]. Combining (f) and (h) yields a special case of Peterson's composition theorem, [**89**].

4.2.5. Sums and linear combinations (cf. [21] exposé 14). The results stated here make use of the notation

$$\Theta(\varphi, \theta)$$

to denote a secondary operation based on a relation $[\varphi][\theta] = 0$ where

$$C_0 \xrightarrow{\theta} C_1 \xrightarrow{\varphi} C_2\ ,$$

and the tethering, while fixed by Θ, is not otherwise specified. The details are left as an exercise.

(a): If $\beta : C_2 \to D$, then

$$\Theta(\beta\varphi, \theta) + \delta = \Omega\beta\Theta(\varphi, \theta)$$

as natural transformations

$$S_\theta(\ \ ) \to T_{\Omega(\beta\varphi)}(\ \ )$$

and $\delta : C_0 \to \Omega D$. In particular, there is no implied relationship between tetherings on either side of the equation.

(b): If $\alpha : A \to C_0$, then

$$\Theta(\varphi, \theta\alpha) + \delta = \Theta(\varphi, \theta) \circ \alpha$$

as natural transformations

$$S_{\theta\alpha}(\ \ ) \to T_{\Omega\varphi}(\ \ )$$

and $\delta : A \to \Omega C_2$.

(c): If $\psi : C_2 \to C_3$ and $[\psi][\varphi][\theta] = 0$, then

$$\Theta(\psi\varphi, \theta) = \Theta(\psi, \varphi\theta) + \delta$$

as natural transformations

$$S_\theta(\ \ ) \to T_{\Omega\psi}(\ \ )$$

and $\delta : C_0 \to \Omega C_3$. In a,b,c, if α, β, or φ are equivalences, then $\delta = 0$ is possible by a suitable choice of tethering.

(d): If $\theta : C_0 \to C_1$, $\varphi_i : C_1 \to C_2$ $1 \leq i \leq a$ and $[\varphi_i][\theta] = 0$ for each i, then

$$\Theta\left(\sum_i \varphi_i, \theta\right) + \delta = \sum_i \Theta(\varphi_i, \theta)$$

as natural transformations from

$$S_\theta(\ \ ) \text{ to } [-, \Omega C_2]/\sum_i Ind(\Theta_i, -)$$

and $\delta : C_0 \to \Omega C_2$. Here sums are formed from an assumed H-structure on C_2.

(e): If $\theta_i : C_0 \to C_1$, $1 \leq i \leq a$, $\varphi : C_1 \to C_2$ and $[\varphi][\theta_i] = 0$ for each i, then

$$\Theta(\varphi, \sum_i \theta_i) + \delta = \sum_i \Theta(\varphi, \theta_i)$$

as natural transformations from

$$\bigcap_i S_{\theta_i}(\ \ ) \to T_{\Omega\varphi}(\ \ )$$

and $\delta : C_0 \to \Omega C_2$. The sum is formed from an assumed H-structure on C_1.

(f): Another form of linear combination occurs when we are presented with a sequence

$$C_0 \xrightarrow{\theta} C_1 \xrightarrow{\varphi} C_2 \xrightarrow{\psi} C_3$$

where $[\varphi][\theta] = 0$ and the maps are represented by matrices. We work with an example,

$$K(Z,n) \xrightarrow[\binom{Sq^2}{Sq^4}]{} K(n+2,n+4) \xrightarrow[\begin{pmatrix} Sq^2,0 \\ Sq^2Sq^1,Sq^1 \end{pmatrix}]{} K(n+4,n+5) \xrightarrow[(Sq^2,Sq^1)]{} K(n+6) \ .$$

We have projection maps

$$\rho_1, \rho_2 : K(n+4,n+5) \longrightarrow K(n+4), K(n+5) \text{ respectively.}$$

We have the inclusion of factors

$$j_1, j_2 : K(n+4), K(n+5) \longrightarrow K(n+4,n+5) \text{ respectively.}$$

We use the projection maps to construct operations from the rows of φ. Let

$$\Theta_1 = \Theta(\rho_1\varphi, \theta) \ ,$$
$$\Theta_2 = \Theta(\rho_2\varphi, \theta) \ .$$

Set $\psi_1 = \psi \circ j_1$, $\psi_2 = \psi \circ j_2$. Then

$$\Omega\psi\Theta(\varphi,\theta) = \Omega\psi_1\Theta_1 + \Omega\psi_2\Theta_2 + \Omega\psi \circ \delta$$

as natural transformations

$$S_\theta(\ \ ) \longrightarrow [-, \Omega C_3]/\Omega\psi_1 Ind(\Theta_1,-) + \Omega\psi_2 Ind(\Theta_2,-)$$

and $\delta : C_0 \to \Omega C_2$. The key step is the factorization of the identity map on the (H-space) C_2, with its usual product structure,

$$Id \simeq j_1 \circ \rho_1 + j_2 \circ \rho_2 \ .$$

(g): We use the inclusion of factors to construct operations from the columns of φ. This construction is more delicate than the construction based on rows of φ. Assume now that $[\psi][\varphi] = 0$ in the set-up of (f).

Write

$$\varphi = (\varphi_1|\varphi_2)$$

where the entries are the columns of φ. Then, for example

$$\varphi \circ j_1 = \varphi_1$$

and we may write the homotopy commutative diagram

$$\begin{array}{ccc} K(n+2) & \xrightarrow{\varphi_1} & K(n+4,n+5) \\ {\scriptstyle j_1}\downarrow & & \| \\ K(n+2,n+4) & \xrightarrow[\varphi]{} & K(n+4,n+5)\ . \end{array}$$

Thus, we can apply (b) in the situation

$$\Theta(\psi,\varphi_1) = \Theta(\psi,\varphi\circ j_1) = \Theta(\psi,\varphi)\circ j_1 + \delta\ .$$

By naturality of pull-backs, we may regard the space in the upper left corner of the diagram

$$\begin{array}{ccc} W_{Sq^4} & \xrightarrow{Sq^2 \circ p} & K(n+2) \\ {\scriptstyle p}\downarrow & & \downarrow{\scriptstyle j_1} \\ K(Z,n) & \xrightarrow[\theta]{} & K(n+2,n+4) \end{array}$$

as either the pull-back of j_1 over θ or the homotopy fiber of $\rho_2\circ\theta$. Then $\rho_1\circ\theta$ determines the map across the top. Thus, selection of the first column of φ leads to a sequence

$$W_{Sq^4} \stackrel{Sq^2\circ p}{\longrightarrow} K(n+2) \stackrel{\varphi_1}{\longrightarrow} C_2 \stackrel{\psi}{\longrightarrow} C_3$$

with

$$[\psi][\varphi_1] = 0$$

and

$$[\varphi_1][Sq^2\circ p] = 0\ .$$

The best information in this set-up is usually obtained by using the methods of Chapter 7 to study W, and comes in the form of

$$\Omega\psi\circ\Theta(\varphi_1, Sq^2\circ p)\ .$$

Exercise 4.2.5. Detection of η^3. We outline the use of the present methods for this composition. This work anticipates section 3.

Let Θ be a secondary operation based on

$$Sq^2Sq^2 = 0$$

for

$$K(Z,n) \stackrel{Sq^2}{\longrightarrow} K(Z/2,n+2) \stackrel{Sq^2}{\longrightarrow} K(Z/2,n+4),\ n\geq 3\ .$$

The fact that Θ is non-trivial on the mapping cone of η^2,

$$T_{\eta^2} = S^n\cup_{\eta^2} e^{n+3}$$

has been established. Here we shall find a "relation" for

$$Sq^2\circ\Theta$$

and use it in conjunction with (4.2.5)(e) to prove that η^3 is essential. We begin with the observation that each composite pair in the following sequence is trivial,

$$K(Z,n) \xrightarrow{\binom{Sq^2}{Sq^4}} K(Z/2,n+2)\times K(Z/2,n+4) \xrightarrow{\begin{pmatrix} Sq^2,0 \\ Sq^2Sq^1,Sq^1 \end{pmatrix}}$$
$$K(Z/2,n+4)\times K(Z/2,n+5) \xrightarrow{(Sq^2,Sq^1)} K(Z/2,n+6)\ .$$

We represent this information abstractly as

$$C_0 \xrightarrow{\theta} C_1 \xrightarrow{\varphi} C_2 \xrightarrow{\psi} C_3\ .$$

Note that $\varphi\theta$ contains the relation for Θ. If we write Φ for a secondary operation based on the other component in the matrix representation for $\varphi\theta$,

$$(Sq^2Sq^1)Sq^2 + Sq^1Sq^4 \equiv 0(Sq^1)\ .$$

Then, we have a universal example for

$$Sq^2\Theta + Sq^1\Phi$$

displayed below

$$\begin{array}{ccccc} W_\theta & \xrightarrow{\tilde{\varphi}} & \Omega C_1 & \xrightarrow{\Omega\psi} & \Omega C_3 \\ \downarrow{\scriptstyle p} & & & & \\ C_0\ . & & & & \end{array}$$

Lemma 4.2.6. $\Omega\psi \circ \tilde{\varphi}$ is null, for any choice of $\tilde{\varphi}$.

Proof. Since $\varphi\theta$ and $\psi\varphi$ are both null-homotopic, there is a factorization

$$\Omega\psi \circ \tilde{\varphi} = \delta \circ p$$

for some $\delta : C_0 \to \Omega C_3$. Now $H^{n+5}(C_0) = H^{n+5}(K(Z,n);Z/2)$ is one of the following:

$$\begin{aligned} &n \geq 5\ \text{,span}\ \{Sq^5 b_n\}\ , \\ &n = 4\ ,0\ , \\ &n = 3\ \text{,span}\ \{b_3 \cup Sq^2 b_3\}\ . \end{aligned}$$

It follows from (3.3.5), that $p \circ \delta$ is null-homotopic.

Now, we can detect η^3 with an operation based on the null-composition $\Omega\psi \circ \tilde{\varphi}$,

$$\mathbb{T} : S_{\tilde{\varphi}}(\quad) \to T_{\Omega^2\psi}(\quad)$$

associated to

$$W_\theta \xrightarrow{\tilde{\varphi}} \Omega C_2 \xrightarrow{\Omega\psi} \Omega C_3\ .$$

We use the factorization $(\eta^2)(\eta)$ to place the mapping cone of η^3 in a cofibration sequence,

$$\begin{array}{ccccc}
S^{n+3} & = & S^{n+3} & & \\
\downarrow{\scriptstyle \eta} & & \downarrow{\scriptstyle \eta^3} & & \\
S^{n+2} & \xrightarrow{\eta^2} & S^n & \longrightarrow & S_{\eta^2} \\
\downarrow & & \downarrow & & \| \\
T_\eta & \xrightarrow{f} & T_{\eta^3} & \xrightarrow{j} & T_{\eta^2} \, .
\end{array}$$

The switchback portion of the calculation of Δ,

$$\Delta = \Delta(\Omega\psi, \tilde{\varphi}) \, ,$$

is displayed below:

$$\begin{array}{ccccccc}
[T_{\eta^2}, W_\theta] & \xrightarrow{\tilde{\varphi}_\#} & [T_{\eta^2}, \Omega C_2] & \xleftarrow{q^\#} & [\Sigma T_\eta, \Omega C_2] & \xrightarrow{\Omega\psi_\#} & [\Sigma T_\eta, \Omega C_3] \\
\downarrow{\scriptstyle surj} & & \downarrow{\scriptstyle \cong} & & & & \\
S_\theta(T_{\eta^2}) & = & H^n(T_{\eta^2}, Z) & \xrightarrow[T_{\Omega\varphi}(T_{\eta^2})]{} & T_{\Omega\varphi}(T_{\eta^2}) \, . & &
\end{array}$$

Here

$$[T_{\eta^2}, \Omega C_2] = H^{n+3}(T_{\eta^2}) \oplus H^{n+4}(T_{\eta^2}) \, ,$$
$$[\Sigma T_\eta, \Omega C_2] = H^{n+3}(\Sigma T_\eta) \oplus H^{n+4}(\Sigma T_y)$$

with both second summands equal to zero. Furthermore, $q^\#$ is an isomorphism of $H^{n+3} \neq 0$. Then $\Omega\psi_\#$ acts via Sq^2 and is non-zero.

The source and target for $\Delta = \Delta(\Omega\psi, \tilde{\varphi})$ are determined as follows:

$$A : [T_{\eta^2}, W_\theta] \longrightarrow [T_{\eta^3}, \Omega C_2]$$

is the zero map, because $A = j^\# \circ \tilde{\varphi}_\#$ and $j^\# = 0$ on

$$[T_{\eta^2}, \Omega C_2]$$

and

$$B : [\Sigma T_{\eta^3}, \Omega C_2] \longrightarrow [\Sigma T_\eta, \Omega C_3]$$

is the zero map because the source is 0. Hence

$$\ker A = [T_{\eta^2}, W_\theta], \mathrm{coker} B = [\Sigma T_\eta, \Omega C_3] \, .$$

The element ϵ representing a generator of

$$H^n(T_{\eta^3}, Z)$$

is in the image of

$$j^{\#} : \ker A \to S_{\tilde{\varphi}}(T_{\eta^3}) \; .$$

Finally

$$\begin{array}{ccc} (\Sigma f)^{\#} : T_{\Omega\psi}(\Sigma T_{\eta^3}) & \longrightarrow & [\Sigma T_{\eta}, \Omega C_3] \\ \| & & \| \\ H^{n+5}(\Sigma T_{\eta^3}) & \longrightarrow & H^{n+5}(\Sigma T_{\eta}) \end{array}$$

is an isomorphism. Thus we have constructed and calculated a natural transformation

$$\mathbb{T} : S_{\tilde{\varphi}}(T_{\eta^3}) \longrightarrow T_{\Omega^2\psi}(T_{\eta^3}) \; .$$

We now apply our calculation to show that η^3 is essential. If not, there is a homotopy equivalence

$$T_{\eta^3} \simeq S^n \vee S^{n+4}$$

and a map

$$T_{\eta^3} \xrightarrow{ret} S^n \xrightarrow{\epsilon_1} C_0$$

with ϵ_1 generating $S_\theta(S^n)$. Let $\bar{\epsilon}_1$ be a lift of ϵ_1

$$S^n \xrightarrow{\bar{\epsilon}_1} W_\theta \; .$$

Now $\bar{\epsilon}_1$ is in $S_{\tilde{\varphi}}(S^n)$ and $T(\bar{\epsilon}_1) = 0$, both for dimensional reasons. From our previous calculation, we have a composition

$$\epsilon : T_{\eta^3} \xrightarrow{j} T_{\eta^2} \longrightarrow W_\theta$$

such that $\mathbb{T}(\epsilon) \neq 0$. The difference between

$$\bar{\epsilon}_1 \circ ret \text{ and } \epsilon$$

is measured by $[T_{\eta^3}, \Omega C_1] = 0$. Thus the operation $\mathbb{T}$ distinguishes T_{η^3} from $S^n \vee S^{n+4}$ and shows that η^3 is essential. □

We will use this calculation to help fix ideas in a discussion of "tertiary" operations to follow. Here we take note of an example where $\mathbb{T}$ fails to distinguish homotopy types. Let Z be obtained by attaching a cell of dimension $(n+4)$ to the bouquet

$$Z = (S^n \vee S^{n+1}) \cup_\gamma e^{n+4}$$

where $\gamma = (\eta^3, \eta^2)$. There is a map

$$f : Z \to T_{\eta^3}$$

obtained by pinching S^{n+1} to a point in Z. Furthermore,

$$\mathbb{T}(\epsilon \circ f) \neq 0 \text{ in } T_{\Omega^2\psi}(Z) \; .$$

However, $[Z, \Omega C_1] \neq 0$, so it is possible that

$$\epsilon \circ f \neq \bar{\epsilon}_1 \circ (ret) \circ f ,$$

were Z to have S^n as a retract. The issue is the indeterminacy.

4.2.6. Stable secondary operations. Let $\{\theta_m\}$, $\{\varphi_m\} m \geq 0$ denote a pair of stable primary operations with components

$$\theta_m : C_{0,m} \longrightarrow C_{1,m} ,$$
$$\varphi_m : C_{1,m} \longrightarrow C_{2,m} ,$$

and identifications $\Omega C_{i,m+1} \simeq C_{i,m}$, $m \geq 0$, $0 \leq i \leq 2$, such that

$$\Omega\theta_{m+1} \simeq \theta_m, \ \Omega\varphi_{m+1} \simeq \varphi_m .$$

Suppose $[\varphi_m][\theta_m] = 0$ for $m \geq 0$. We wish to use this data to construct a sequence of secondary operations,

$$\{\Theta_m\} , \ \Theta_m : S_{\theta_m}(\ \ ) \to T_{\Omega\varphi_m}(\ \ )$$

which enjoy both compatibility under looping, and after a suitable sign adjustment, compatibility with exact sequences.

We introduce the following modification in the definition of Δ from subsection (4.2.3). Write

$$\tau_m = \begin{Bmatrix} \text{identity map} & m \text{ even} \\ \tau & m \text{ odd} \end{Bmatrix}$$

where τ is the direction reversal map. Set

$$\Delta_m = (-1)^m \Delta(\varphi_m, \theta_m) = \Delta(\varphi_m, \tau_m \circ \theta_m) .$$

Definition 4.2.7. A sequence of secondary operations

$$\{\Theta_m\}, \ \Theta_m : S_{\theta_m}(\ \ ) \to T_{\Omega\varphi_m}(\ \ )$$

for $m \geq 0$ is <u>stable</u> provided

- **(a):** $\Omega\Theta_{m+1} = \Theta_m$;
- **(b):** Θ_m is compatible with Δ_m, in the sense of Prop. 4.2.2, with φ_m, θ_m substituted for φ, θ respectively.

Before discussing the existence of stable secondary operations, we look at an example. We have the Adem relation

$$[\varphi][\theta] = 0$$

where

$$[\varphi] = (2\mathcal{P}^2, \beta\mathcal{P}^1 - 2\mathcal{P}^1\beta) ,$$

$$[\theta] = \begin{pmatrix} \beta \\ \mathcal{P}^1 \end{pmatrix} ,$$

and β is the connecting operator associated to the short exact sequence

$$0 \to Z/p \to Z/p^2 \to Z/p \to 0 .$$

To obtain stable primary operations, we replace β by the signed Bockstein

$$\mathfrak{B} = (-1)^n \beta, \text{ for } \beta : H^n(\, ; Z/p) \to H^{n+1}(\, ; Z/p) .$$

The sign is determined by the dimension of the domain.

To construct a stable secondary operation, we take

$$\begin{aligned} C_{0,m} &= K(Z/p, m+1) , \\ C_{1,m} &= K(Z/p, m+2) \times K(Z/p, m+2p-1) , \\ C_{2,m} &= K(Z/p, m+4p-2) \quad m \geq 0 . \end{aligned}$$

We write

$$\begin{aligned} \theta_m : &C_{0,m} \to C_{1,m} \text{ given by } \begin{pmatrix} \mathfrak{B} \\ \mathcal{P}^1 \end{pmatrix} , \\ \varphi_m : &C_{1,m} \to C_{2,m} \text{ given by } (2\mathcal{P}^2, \mathfrak{B}\mathcal{P}^1 - 2\mathcal{P}^1\mathfrak{B}) . \end{aligned}$$

It is now easy to produce an initial segment of components for a stable secondary operation. Choose M to be a large integer. For $m < M$, define k by

$$m = M - k .$$

Let H_M be a contracting homotopy for

$$\varphi_M \circ \theta_M .$$

Let Θ_m, $0 \leq m \leq M$, be the secondary operation associated with

$$(\varphi_m, \tau_m \circ \theta_m, \tau_m \circ \Omega^k H_M) .$$

Then, both parts of Defn. 4.2.7 are satisfied; since

$$\Omega\Theta_M$$

is associated with

$$\begin{aligned} &(\varphi_{M-1}, \tau \circ \Omega(\tau_M \circ \theta_M), \ \tau \circ \Omega(\tau_M \circ H_M)) \\ &= (\varphi_{M-1}, \tau_{M-1} \circ \theta_{M-1}, \ \tau_{M-1} \circ \Omega H_M) \end{aligned}$$

and so forth. Likewise, Θ_m is compatible with Δ_m.

Now we look at the obstruction to extending this construction by one. Let L be a contracting homotopy for the composition

$$\varphi_{M+1} \circ \theta_{M+1} ,$$

and let Θ_{M+1} be associated with

$$(\varphi_{M+1}, \ \tau_{M+1} \circ \theta_{M+1}, \ \tau_{M+1} \circ L) .$$

The difference between $\Omega\Theta_{M+1}$ and Θ_M is given by,

$$\begin{aligned}\delta &= \delta(\tau_M \circ H_M,\ \tau_M \circ \Omega L)\\ &= \delta(H_M, \Omega L) : C_{0,M} \longrightarrow \Omega C_{2,M}\ .\end{aligned}$$

We wish to extend the adjoint of δ through $C_{0,M+1}$, as in the diagram

$$\begin{array}{ccc} \Sigma C_{0,M} & \xrightarrow{\delta^\flat} & \Omega C_{2,M+1} \\ {\scriptstyle j_{1,\infty}}\downarrow & & \| \\ C_{0,M+1} & \dashrightarrow & \Omega C_{2,M+1}\ . \end{array}$$

In general, this extension is possible under the

4.2.7. Stability hypothesis. All maps

$$\Sigma E_k(\Omega C_{0,m+1}) \longrightarrow C_{2,m+1}\ ,$$

for $k \geq 1$ and m sufficiently large, are null-homotopic, where E_k is the space described in Prop. 3.5.1.

In the case of our example,

$$\Sigma E_k(\Omega C_{0,m+1}) = (\Sigma K(Z/p, m+1))^{(K+1)}\ ,$$

which is at least $(2m+3)$-connected. Moreover,

$$C_{2,m+1} = K(Z/p, m+4p-1)$$

and the stability hypothesis holds for

$$M \geq 4p-4\ .$$

The operation constructed in this example is denoted by $\mathcal{R}$.

In summary we have

Proposition 4.2.8. Given a pair of stable primary operations

$$\{\theta_m\}, \{\varphi_m\} \text{ with each } [\varphi_m][\theta_m] = 0\ ,$$

and the stability hypothesis, then there exists a stable secondary operation

$$\{\Theta_m\}\ ,$$

$$\Theta_m : S_{\theta_m}(\quad) \longrightarrow T_{\Omega\varphi_m}(\quad)$$

where each Θ_m is based on the relation

$$[\varphi_m][\tau_m \circ \theta_m] = 0\ .$$

Furthermore, if

$$\{\Theta_m\} \text{ and } \{\Theta'_m\}$$

are two stable secondary operations based on the same relation, then there is a stable primary operation

$$\{\delta_m\}\ ,$$
$$\delta_m : C_{0,m} \longrightarrow \Omega C_{2,m}$$

such that

$$\Theta'_m - \Theta_m = \delta_m\ .$$

Exercise. Let p be an odd prime. Each pair of compositions in the sequence

$$S^{4p-3} \xrightarrow{\alpha_1(2p)} S^{2p} \xrightarrow{\times p} S^{2p} \xrightarrow{\alpha_1(3)} S^3$$

is null (e.g. by the exponent theorem [**98**]). For each $n \geq 3$, we can form secondary compositions

$$C : S^{n+4p-5} \underset{a_2}{\longrightarrow} P^{n+2p-2}(p) \underset{a_1}{\longrightarrow} S^n$$

where a_1 is an extension of $\alpha_1(n)$ and a_2 is a coextension of $\alpha_1(n+2p-3)$. Prove that a_1 and a_2 are detected by a component of the factors $[\theta], [\varphi]$ respectively, of $\mathcal{R}$, and that a component of secondary operation $\mathcal{R}$ detects C.

4.2.8. Algebraic data for stable secondary operations. In the classical situation, stable operations are often constructed in the following manner. Let X_0, X_1 be finitely generated, free, graded $\mathcal{A}$-modules. We shall produce our operation from homogeneous d-cycles, where

$$d : X_1 \to X_0$$

is an $\mathcal{A}$-map. Write the generators of X_0, X_1 respectively as

$$x_{0,1}, \ldots, x_{0,R}\ ,$$
$$x_{1,1}, \ldots, x_{1,S}\ .$$

Define a map d by the equation, with homogeneous terms,

$$d(x_{1,s}) = \sum_r a_{s,r} x_{0,r}\ .$$

Let z be a homogeneous d-cycle

$$z = \Sigma b_s x_{1,s}\ .$$

Let φ, θ be the following stable primary operations,

$$\varphi = (\tilde{b}_1, \ldots, \tilde{b}_S) \quad 1 \times S \text{ matrix}\ ,$$
$$\theta = (\tilde{a}_{s,r}) \qquad S \times R \text{ matrix},$$

where the tilde indicates the stable operation corresponding to an element of the Steenrod algebra.

Then

$$\varphi\theta = 0 .$$

Furthermore, the stability hypothesis holds for finite products of Eilenberg-Mac Lane spaces. Hence we may apply Prop. 4.2.8.

Signs again.

Our application of Prop. 4.2.8 uses the element $a_{s,r}$ in $\mathcal{A}$ to determine a stable primary operation by introducing the signed Bockstein, in the case of odd primes. In practice, one may prefer to work directly with the ordinary Bockstein. The following discussion provides a means for handling the resulting signs.

In the formalism of subheading (1.3.6) we have

$$\theta : C_{0,m} \longrightarrow C_{1,m} \text{ represented by } (a_{s,r}) ,$$
$$\varphi : C_{1,m} \longrightarrow C_{2,m} \text{ represented by } (b_s) .$$

The relation $[\varphi][\theta] = 0$ yields, for each m, a secondary operation Θ_m (based on $[\varphi][\tau_m \circ \theta]$). In particular,

$$\Omega\Theta_{m+1} \text{ is based on } [\Omega\varphi][\tau_m \circ \Omega\theta]$$

and

$$\Omega\theta \text{ is represented by } ((-1)^{|a_{s,r}|}a_{s,r}) ,$$
$$\Omega\varphi \text{ is represented by } ((-1)^{|b_s|}b_s) .$$

To determine the signs involved in comparing Θ_m with $\Omega\Theta_{m+1}$, we introduce some notation. Consider an n-tuple of integers

$$x = (x_1, \ldots, x_n) .$$

Write

$$(-1)^x = ((-1)^{x_1}, \ldots, (-1)^{x_n})$$

for the n-tuple of ± 1 depending on the parity of the components of x. Write

$$\operatorname{diag}(-1)^x = \begin{pmatrix} (-1)^{x_1} & & 0 \\ & \ddots & \\ 0 & & (-1)^{x_n} \end{pmatrix}$$

for the $n \times n$ diagonal matrix with $(-1)^{x_i}$ in the i-th diagonal position. Given two n-tuples x, y, we say x is <u>similar</u> to y if

$$(-1)^x = (-1)^y ,$$

and <u>complementary</u> if

$$(-1)^x = -(-y)^y .$$

Thus, in the definition, the parities of the components of x and y either agree or disagree, independently of their position in the n-tuple. If either of the conditions holds, we say that x and y are <u>compatible</u>.

Next we look at the top row and left column of the matrix $(a_{s,r})$ representing θ. Write TR for the R-tuple with components equal to the degrees of the elements in the top row,

$$TR = (|a_{1,1}|, \ldots, |a_{1,R}|) .$$

We understand $|x_{1,1}| - |x_{0,r}|$ in place of $|a_{1,r}|$ if $a_{1,r} = 0$. Likewise for the left column,

$$LC = (|a_{1,1}|, \ldots, |a_{S,1}|) .$$

We do the same for the matrix (= row in this case) representing φ,

$$|b| = (|b_1|, \ldots, |b_S|) .$$

Lemma 4.2.9. The rows of $(a_{s,r})$ are compatible with TR.

The columns of $(a_{s,r})$ and $|b|$ are compatible with LC. Moreover, we have the equations,

$$(a_{s,r}) = (\mathrm{diag}(-1)^{LC})((-1)^{|a_{s,r}|}a_{s,r})((-1)^{|a_{1,1}|}\mathrm{diag}(-1)^{TR})$$

and

$$(b_s) = (-1)^{|z|-|x_{0,1}|}((-1)^{|b_s|}b_s)(\mathrm{diag}(-1)^{LC}) .$$

Proof. Consideration of degrees yields the equations

$$|a_{s,r}| + |x_{0,r}| = |a_{s,1}| + |x_{0,1}| ,$$
$$|a_{1,r}| + |x_{0,r}| = |a_{1,1}| + |x_{0,1}| .$$

Thus

$$|a_{s,r}| = |a_{s,1}| + |a_{1,r}| - |a_{1,1}| .$$

This equation establishes row and column compatibility for $(a_{s,r})$ and the first matrix equation. Similarly, we have

$$|z| = |b_s| + |a_{s,1}| + |x_{0,1}| .$$

The remaining statement follows from this equation. □

Thus, in general, we have

$$\Theta_m = (-1)^{|z|-|x_{0,1}|}\Omega\Theta_{m+1}\left((-1)^{|a_{1,1}|}\mathrm{diag}(-1)^{TR}\right) .$$

Next, we specialize to the case where Θ_m is a function of one variable. Here $R = 1$ and likewise for the diagonal matrix on display. Moreover, take $|x_{0,1}| = 0$ as a normalizing condition. Then we have

$$\Theta_m = (-1)^{|z|}\Omega\Theta_{m+1} .$$

Next we turn to relations involving primary and secondary operations, of which exercise (4.2.5) supplies an example. In terms of our algebraic data, we have homogeneous d-cycles

$$z_t = \sum_s b_{t,s} x_{1,s} \qquad 1 \le t \le T \, .$$

Let X_2 be a free $\mathcal{A}$-module on T generators

$$x_{2,1}, \ldots, x_{2,T}$$

and define

$$d_1 : X_2 \to X_1$$

by

$$d_1(x_{2,t}) = z_t = \sum_s b_{t,s} x_{1,s} \, .$$

We are interested in homogeneous elements in kernel d_1,

$$w = \Sigma c_t x_{2,t} \text{ such that } \Sigma c_t z_t = 0 \, .$$

We now have maps

$$\begin{aligned} \theta &: C_{0,m} \longrightarrow C_{1,m} \text{ represented by } (a_{s,r}) \, , \\ \varphi &: C_{1,m} \longrightarrow C_{2,m} \text{ represented by } (b_{t,s}) \, , \\ \psi &: C_{2,m} \longrightarrow C_{3,m} \text{ represented by } (c_t) \, , \end{aligned}$$

and each composite

$$\psi \circ \varphi, \; \varphi \circ \theta$$

is null-homotopic.

This data is used to construct universal examples on display below:

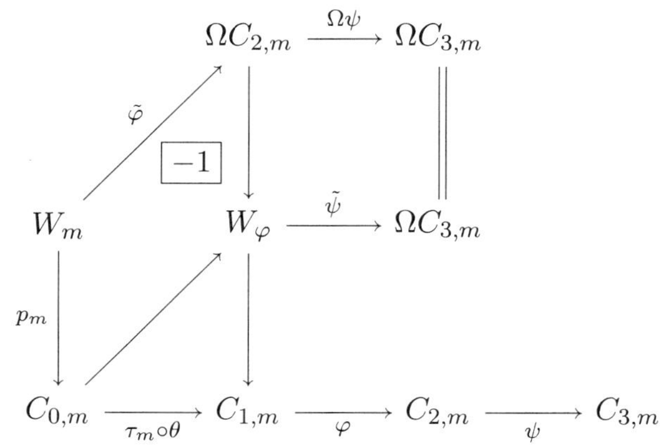

where W_m is an abbreviated notion for $W_{\tau_m \circ \theta}$. Note that

$$\Omega\psi \text{ is represented by } ((-1)^{|c_t|} c_t)$$

independently of m. We write $\Theta_{m,t}$ for the secondary operation based on $[{}_t\varphi][\tau_m \circ \theta]$ where ${}_t\varphi$ denotes the t-th row of φ.

From the diagram, we have

$$\Omega\psi \circ \tilde{\varphi} = \delta_m \circ p_m, \quad \delta_m : C_{0,m} \to \Omega C_{3,m} .$$

We write this equation in terms of operations, using subsection (4.2.5)(f), and obtain

$$\sum_t (-1)^{|c_t|} c_t \Theta_{m,t}(p_m) = [\![\delta_m \circ p_m]\!] .$$

As before, we may wish to use this equation in conjunction with looping. With $R = 1$ and $|x_{0,1}| = 0$, we have

$$\Omega(\Sigma(-1)^{|c_t|} c_t \Theta_{m+1,t}(p_{m+1})) = (-1)^{|w|}(\Sigma(-1)^{|c_t|} c_t \Theta_{m,t}(p_m))$$

and

$$[\![\Omega(\delta_{m+1} \circ p_{m+1})]\!] = [\![\Omega\delta_{m+1} \circ p_m]\!] .$$

This formula follows from the formula for the summands and the equation

$$|w| = |c_t| + |z_t| .$$

In many cases, there is also a stable operation $\delta = (\delta_m)$, obtained, for example, by compatible choices of tetherings, or in cases where the results are independent of tetherings. □

4.2.9. Measure of non-additivity. Suppose we have a sequence with homotopy

$$(\Omega\varphi, \Omega\theta, H) ,$$

$$\Omega C_0 \xrightarrow{\Omega\theta} \Omega C_1 \xrightarrow{\Omega\varphi} \Omega C_2$$

but $[\varphi][\theta] \neq 0$. We use the discussion in subsection (3.5.7) to obtain a geometric measure of the failure of Θ to be additive, for a secondary operation based on $[\Omega\varphi][\tau \circ \Omega\theta] = 0$,

$$\Theta : S_{\Omega\theta}(\ \ ) \to T_{\Omega^2\varphi}(\ \ ) .$$

Take $f = \varphi\theta$. Then φ induces a canonical map

$$h : W_\theta \to W_f$$

from the pull-back construction

$$\begin{array}{ccccc} C_0 & \xrightarrow{\theta} & C_1 & \longleftarrow & P_0C_1 \\ \| & & \downarrow{\scriptstyle\varphi} & & \downarrow{\scriptstyle P_0\varphi} \\ C_0 & \xrightarrow[f]{} & C_2 & \longleftarrow & P_0C_2 . \end{array}$$

Moreover, $\Omega\varphi$ induces a map

$$h' : W_{\tau\circ\Omega\theta} \longrightarrow W_{\tau\circ\Omega f}$$

from the pull-back construction

$$\begin{array}{ccccc} \Omega C_0 & \xrightarrow{\tau\circ\Omega\theta} & \Omega C_1 & \longleftarrow & P_0\Omega C_1 \\ \Vert & & \downarrow{\scriptstyle \Omega\varphi} & & \downarrow{\scriptstyle P_0\Omega\varphi} \\ \Omega C_0 & \xrightarrow[\tau\circ\Omega f]{} & \Omega C_2 & \longleftarrow & P_0\Omega C_2\ . \end{array}$$

We define a colifting, by composition,

$$\widetilde{\Omega\varphi} : W_{\tau\circ\Omega\theta} \xrightarrow{h'} W_{\tau\circ\Omega f} \xrightarrow{\tilde{I}} \Omega^2 C_2$$

where $\tilde{I}$ is from Prop. 3.5.5 and we use $L : I \times \Sigma\Omega C_0 \to C_2$ given by $L(t,s,w) = H(t,w)(s)$. Then the following diagram commutes up to homotopy:

$$\begin{array}{ccc} W_{\tau\circ\Omega\theta} & \xleftarrow{\rho_\theta} & \Omega W_\theta \\ \downarrow{\scriptstyle \widetilde{\Omega\varphi}} & & \downarrow{\scriptstyle \gamma^{\natural\natural}\circ\Omega h} \\ \Omega^2 C_2 & \xleftarrow[\kappa\circ\tau]{} & \Omega^2 C_2 \end{array}$$

This is immediate from Prop. 3.5.5 and the observation that

$$\rho_f \circ \Omega h \simeq h' \circ \rho_\theta\ .$$

Besides containing useful information for calculations, the square above allows us to replace the universal example for Θ associated with $(\Omega\varphi, \tau\circ\Omega\theta, \tau\circ H)$ with

$$\Omega W_\theta \xrightarrow{\gamma^{\natural\natural}\circ\Omega h} \Omega^2 C_2\ .$$

The geometric measure of non-additivity for Θ is obtained by juxtaposing this result with the diagram in Prop. 3.5.4. For the main case of interest, $k = 1$, we have

$$\gamma \circ \Sigma^2\Omega h \circ \Sigma p_1 \simeq \zeta \circ (\Sigma\Omega q)^{(2)}$$

and $q : W_\theta \to C_0$ is the projection. Moreover, if $k > 1$, then Θ is an additive operation represented by $\gamma^{\natural\natural} \circ \Omega h$. In fact, we have

$$\gamma^{\natural\natural} : \underbrace{\Omega W_f \longrightarrow \Omega\Sigma\Omega W_f \longrightarrow \Omega B_k W_f}_{H\text{-map}} \xrightarrow{\Omega(\gamma^{\natural})} \Omega^2 C_2, k \geq 2$$

For $k = 1$ and ϵ_1, ϵ_2 in $S_{\Omega\theta}(X)$, we have

$$\Theta(\epsilon_1 + \epsilon_2) = \Theta(\epsilon_1) + \Theta(\epsilon_2) + [\![\zeta^{\natural\natural} \circ \epsilon_1 \wedge \epsilon_2 \circ \bar{\Delta}]\!]\ .$$

Thus the map ζ in Prop. 3.5.4 measures the non-additivity of Θ. It can happen that $\zeta \neq 0$ but Θ is additive, for example, by absorption into the indeterminancy. An example is found in [**31**].

4.3. Higher order operations

We introduce this notion in a general way, in order to convey the idea. Serious calculations are generally made in the presence of extra structure. In our treatment, this structure consists of "admissible syzygys" and "buttrusses".

4.3.1. Data for higher order operations. At the minimum, we can define higher order operations in terms of the following data:

(a): a principal fibration $F \xrightarrow{j} E \xrightarrow{p} B$ with the principal action

$$\mu : F \times E \to E ,$$

and

(b): a map $\varphi : E \to D$.

Then, for any space X, we have

$$S_p(X) = \{[\epsilon] | \epsilon : X \to B,\ \epsilon \simeq p \circ \bar{\epsilon} \text{ for some } \bar{\epsilon} : X \to E\}$$

and

$$T_\varphi(X) = im\varphi_\#/R$$

where R is the equivalence relation induced by μ. Thus, for maps

$$\epsilon_1, \epsilon_2 : X \to E,$$

their compositions with φ are R-equivalent provided there is a map

$$\delta : X \to F$$

such that, up to homotopy, $\varphi \circ \epsilon_2$ is the composition

$$X \xrightarrow{\Delta} X \times X \xrightarrow{\delta \times \epsilon_1} F \times E \xrightarrow{\mu} E \xrightarrow{\varphi} D .$$

In this formulation of an operation, the point of view shifts to the universal example. In the course of this work, the role of φ has been taken by coliftings, and we have $\varphi \circ \mu$ factoring through a product, as in the diagram

$$\begin{array}{ccc} F \times E & \xrightarrow{\mu} & E \\ \Big\downarrow{\scriptstyle \varphi \circ j \times \varphi} & & \Big\downarrow{\scriptstyle \varphi} \\ D \times D & \xrightarrow[\mu_D]{} & D , \end{array}$$

where μ_D is a loop space multiplication on D and $T_\varphi(X)$ is replaced by a quotient of $[X, D]$ determined by an action $F \times D \to D$.

Definition 4.3.1. A higher order operation is a natural transformation

$$\mathcal{U} : S_p(\ \ ) \to T_\varphi(\ \ )\,.$$

A generic example is given by

$$\mathcal{U}(\epsilon) = [\![\varphi \circ p^{-1} \circ \epsilon]\!] \text{ in } T_\varphi(X).$$

This generality does not embrace a useful theory. In practice, universal examples for operations are built inductively through towers of fibrations. The data for the construction is

Definition 4.3.2. A syzygy of length n, $n \geq 2$, is a sequence of maps and spaces

$$C_0 \xrightarrow{\theta_0} C_1 \xrightarrow{\theta_1} \cdots \xrightarrow{\theta_{n-1}} C_n$$

such that each composition

$$\theta_{j+1} \circ \theta_j,\ 0 \leq j \leq n-2$$

is null-homotopic, and C_i is i-connected. A syzygy of length n is admissible if either $n = 2$, or if $n > 2$ and there is a contracting homotopy for the composition

$$\theta_1 \circ \theta_0$$

such that

$$W_{\theta_0} \xrightarrow{\tilde{\theta}_1} \Omega C_2 \xrightarrow{\Omega\theta_2} \Omega C_3 \longrightarrow \cdots \xrightarrow{\Omega\theta_{n-1}} \Omega C_n$$

is an admissible syzygy of length $n-1$.

The term "order", in use here, does not have an invariant status. But we think of the length of a syzygy as a measure of imposed order.

4.3.2. Towers. An admissible syzygy of length n yields a tower of "height" n, together with a map from the top space in the tower to $\Omega^{n-1}C_n$. We display the tower for the case $n = 4$ (= means by definition)

$$\begin{array}{ccccc}
W_{\widetilde{\Omega\theta_2}} & = & W_2 & \xrightarrow{\widetilde{\Omega^2\theta_3}} & \Omega^3 C_4 \\
 & & \Big\downarrow{\scriptstyle p_2} & & \\
W_{\tilde{\theta}_1} & = & W_1 & \xrightarrow{\widetilde{\Omega\theta_2}} & \Omega C_3 \\
 & & \Big\downarrow{\scriptstyle p_1} & & \\
W_{\theta_0} & = & W_0 & \xrightarrow{\tilde{\theta}_1} & \Omega C_2 \\
 & & \Big\downarrow{\scriptstyle p_0} & & \\
 & & C_0 & \xrightarrow{\theta_0} & C_1 ,
\end{array}$$

where the successive coliftings are obtained from the admissibility condition.

Our theory for secondary operations applies to each "two-stage towerlet." For example, there is the portion of the tower involving the map p_2,

$$\begin{array}{ccccc}
W_2 & \xrightarrow{\widetilde{\Omega^2\theta_3}} & \Omega^3 C_4 & & \\
\Big\downarrow{\scriptstyle p_2} & & & & \\
W_1 & \xrightarrow{\widetilde{\Omega\theta_2}} & \Omega^2 C_3 & \xrightarrow{\Omega^2\theta_3} & \Omega^2 C_4 .
\end{array}$$

This diagram serves as a universal example for a secondary operation

$$\mathcal{T} : S_{\widetilde{\Omega\theta_2}}(\quad) \to T_{\Omega^3\theta_3}(\quad) .$$

Our next task is to compare the operation $\mathcal{T}$ with the higher order operation

$$\mathcal{U} : S_p(\quad) \to T_{\widetilde{\Omega^2\theta_3}}(\quad),$$

associated with the tower, where $p = p_0 p_1 p_2$. To make this comparison, we wish to describe the universal example for $\mathcal{U}$

$$\begin{array}{ccc}
W_2 & \xrightarrow{\widetilde{\Omega^2\theta_3}} & \Omega^3 C_4 \\
\Big\downarrow{\scriptstyle p} & & \\
C_0 & &
\end{array}$$

as a principal fibration. We begin with the fiber of p.

Lemma 4.3.3. In the diagram below, the squares are pull-backs and $p : E \to B$ is a fibration with fiber F,

$$\begin{array}{ccccc}
F' & \xrightarrow{j'} & E' & \longrightarrow & PD \\
{\scriptstyle q'}\downarrow & & \downarrow{\scriptstyle q} & & \downarrow \\
F & \xrightarrow{j} & E & \xrightarrow{\varphi} & D \\
 & & \downarrow{\scriptstyle p} & & \\
 & & B\,. & &
\end{array}$$

Then F', formed as a pull-back over φj, is homotopy equivalent to the homotopy theoretic fiber of the composition pq.

Proof. We have the diagram

$$\begin{array}{ccccc}
F' & \xrightarrow{q'} & F & \longrightarrow & PB \\
{\scriptstyle j'}\downarrow & & \downarrow{\scriptstyle j} & & \downarrow \\
E' & \xrightarrow[q]{} & E & \xrightarrow[p]{} & B
\end{array}$$

with the right-hand square a pull-back square, up to homotopy. Since the left-hand square is a pull-back square, the composite square is a pull-back up to homotopy. ☐

We apply the lemma to the tower to obtain that the fiber of p is the space over the fiber of $p_0 \circ p_1$, as in the diagram

$$\begin{array}{ccccc}
F_2 & \longrightarrow & W_2 & & \\
\downarrow & & \downarrow{\scriptstyle p_2} & & \\
F_1 & \longrightarrow & W_1 & \longrightarrow & \Omega^2 C_3 \\
 & & \downarrow{\scriptstyle p_0 p_1} & & \\
 & & C_0\,, & &
\end{array}$$

and F_1 is determined inductively by the lemma.

In order to obtain a principal action, we require an additional piece of structure. We continue to treat the case $n = 4$ as the general case, and leave it to the reader to confirm that, after the fact, nothing special has been snuck into the argument.

First we expand the notation to rewrite the tower as

$$\begin{array}{ccccc}
W_2 & = & W(\theta_0, \tilde{\theta}_1, \widetilde{\Omega\theta_2}) & & \\
& & \downarrow & & \\
W_1 & = & W(\theta_0, \tilde{\theta}_1) & \xrightarrow{\widetilde{\Omega\theta_2}} & \Omega^2 C_3 \\
& & \downarrow & & \\
W_0 & = & W(\theta_0) & \xrightarrow{\tilde{\theta}_1} & \Omega C_2 \\
& & \downarrow & & \\
& & C_0 & \xrightarrow{\theta_0} & C_1 \, .
\end{array}$$

Thus, this notation displays the maps used to construct the spaces in the tower and tacitly contains the contracting homotopies needed to define the indicated coliftings. The notation is intended only to indicate where we are in the diagram. It does not signify a homotopy type, as that depends on suppressed data.

If we delete the first space and map from the syzygy in Defn. 4.3.2, we obtain a syzygy of length $(n-1)$ beginning with C_1,

$$C_1 \xrightarrow{\theta_1} C_2 \longrightarrow \cdots \xrightarrow{\theta_{n-1}} C_n \, .$$

We now assume that this syzygy is also admissible. Thus, we may construct a tower over C_1,

$$\begin{array}{ccc}
W(\theta_1, \tilde{\theta}_2) & & \\
\downarrow & & \\
W(\theta_1) & \xrightarrow{\tilde{\theta}_2} & \Omega C_3 \\
\downarrow & & \\
C_1 & \xrightarrow{\theta_1} & C_1
\end{array}$$

with height $(n-1)$. Observe that the following compositions are defined,

$$\begin{array}{ccccc}
W(\theta_0) & \xrightarrow{\tilde{\theta}_1} & \Omega C_2 & \xrightarrow{j} & W(\theta_1) \\
W(\theta_0, \tilde{\theta}_1) & \xrightarrow{\widetilde{\Omega\theta_2}} & \Omega^2 C_3 & \xrightarrow{j} & W(\theta_1, \tilde{\theta}_2)
\end{array}$$

and the diagram below anti-commutes,

$$\begin{array}{ccc} W(\theta_0) & \xrightarrow[\boxed{-1}]{\tilde{\theta}_1} & \Omega C_2 \\ \Big\downarrow{\scriptstyle p_0} & & \Big\downarrow{\scriptstyle j} \\ C_0 & \xrightarrow[L_1(\theta_0)]{} & W(\theta_1) \end{array}$$

where $L_1(\theta_0)$ is the lifting of θ_0 using the same contracting homotopy for $\theta_1 \circ \theta_0$ as is used to construct the colifting $\tilde{\theta}_1$. The map $L_1(\theta_0)$ is automatic, and is the first in a sequence of maps which form the structure we call a "buttruss."

Definition 4.3.4. A *buttruss* for a tower of height n, is a sequence of maps

$$L_i(\theta_0) : C_0 \longrightarrow W(\theta_1, \tilde{\theta}_2, \ldots, \widetilde{\Omega^{i-2}\theta_i}) \; 1 \leq i \leq n-2$$

such that $L_i(\theta_0)$ lifts $L_{i-1}(\theta_0)$, $L_0(\theta_0) = \theta_0$, and the diagram

$$\begin{array}{ccc} W(\theta_0, \tilde{\theta}_1, \ldots, \widetilde{\Omega^{i-2}\theta_{i-1}}) & \xrightarrow{\widetilde{\Omega^{i-1}\theta_i}} & \Omega^i C_{i+1} \\ \Big\downarrow & & \Big\downarrow{\scriptstyle j\circ\tau} \\ C_0 & \xrightarrow[L_i(\theta_0)]{} & W(\theta_1, \tilde{\theta}_2, \ldots, \widetilde{\Omega^{i-2}\theta_i}) \end{array}$$

commutes up to homotopy, where the direction reversal map is inserted on the right.

For example, in the case $n = 4$, we are describing the diagram

$$\begin{array}{ccccc} W(\theta_0, \tilde{\theta}_1) & & \xrightarrow{\qquad\qquad \widetilde{\Omega\theta_2} \qquad\qquad} & & \Omega^2 C_3 \\ \Big\downarrow & & & & \Big| \\ W(\theta_0) & \xrightarrow{\quad \tilde{\theta}_1 \quad} & \Omega C_2 & & \Big| \\ \Big\downarrow & & \Big\downarrow{\scriptstyle j\circ\tau} & & \Big\downarrow{\scriptstyle j\circ\tau} \\ C_0 & \xrightarrow[\theta_0]{} C_1 & & & \Big| \\ \Vert & & & & \Big| \\ C_0 & \xrightarrow[L_1(\theta_0)]{\qquad} & W(\theta_1) & & \Big\downarrow \\ \Vert & & & & \\ C_0 & & \xrightarrow[L_2(\theta_0)]{\qquad\qquad\qquad} & & W(\theta_1, \tilde{\theta}_2)\,. \end{array}$$

The existence of a buttruss does not follow assuming only admissibility of the syzygys starting at C_0 and C_1. We shall not formulate a general result. In Chapter 7, we renew the discussion, where their existence is automatically part of an "Adams resolution".

We turn to the use of buttrusses to describe the relation between the operations $\mathcal{T}$ and $\mathcal{U}$. First, we juxtapose the towers over C_0 and C_1,

$$
\begin{array}{ccccccccc}
\text{col. A} & & \text{col. B} & & & & \text{col. C} & & \\
W(\Omega\theta_1, \widetilde{\Omega\theta_2}) & \rightarrow & W(\theta_0, \tilde{\theta_1}, \widetilde{\Omega\theta_2}) & & & & & & \\
\downarrow & & \downarrow & & & & & & \\
W(\Omega\theta_1) & \longrightarrow & W(\theta_0, \tilde{\theta_1}) & \xrightarrow{\widetilde{\Omega\theta_1}} & \Omega^2 C_3 & \xrightarrow{j} & W(\theta_1, \tilde{\theta_2}) & & \\
\downarrow & & \downarrow & & & & \downarrow & & \\
\Omega C_1 & \longrightarrow & W(\theta_0) & \xrightarrow{\tilde{\theta_1}} & \Omega C_2 & \xrightarrow{j} & W(\theta_1) & \xrightarrow{\tilde{\theta_2}} & \Omega C_3 \\
 & & \downarrow & & & & \downarrow & & \\
 & & C_0 & & \xrightarrow{\theta_0} & & C_1\,. & &
\end{array}
$$

Observe that the spaces in column A are homotopy equivalent to spaces obtained by looping the spaces in column C (and we are careless of signs). The presence of buttresses produces a principal action of the spaces in column A on those in column B. For example, from the diagram

$$
\begin{array}{ccccc}
 & & W(\Omega\theta_1) & \xrightarrow{\simeq} & \Omega W(\theta_1) \\
 & & \downarrow & & \downarrow \\
W(\theta_0, \tilde{\theta_1}, \widetilde{\Omega\theta_2}) & \xrightarrow{p_2} & W(\theta_0, \tilde{\theta_1}) & \xrightarrow{\widetilde{\Omega\theta_2}} & \Omega^2 C_3 \\
\downarrow & & \downarrow{\scriptstyle p \circ p_1} & & \downarrow{\scriptstyle j \circ \tau} \\
W(L_2(\theta_0)) & \longrightarrow & C_0 & \xrightarrow[L_2(\theta_0)]{} & W(\theta_1, \tilde{\theta_2})
\end{array}
$$

we may identify p_2 with the principal fibration induced by $L_2(\theta_0)$.

Hence the secondary operation $\mathcal{T}$ computes $\mathcal{U}$ by

Proposition 4.3.5. $T_{\widetilde{\Omega^2\theta_3}}(\quad)$ is a quotient of $T_{\Omega^3\theta_3}(\quad)$ and the following diagram commutes:

$$\begin{array}{ccccc}
S_{\widetilde{\Omega\theta_2}}(\quad) & \xrightarrow{\mathcal{T}} & T_{\Omega^3\theta_3}(\quad) & \longrightarrow & T_{\widetilde{\Omega^2\theta_3}}(\quad) \\
\Big\downarrow{\scriptstyle p_\#} & & & & \Big\| \\
S_p(\quad) & & \xrightarrow[\mathcal{U}]{\hspace{6em}} & & T_{\widetilde{\Omega^2\theta_3}}(\quad)\,.
\end{array}$$

We can use the methods of this chapter to calculate $\mathcal{T}$. The values of $\mathcal{U}$ lie in a quotient which, roughly, results from dividing by the image of an operation with order lower by one from the order of $\mathcal{U}$. In particular, $\widetilde{\Omega^2\theta_3}$ is compatible with the composite principal action in the sense of subheading (3.2.8).

Remark. The reader can find a use of buttrusses in "Unstable p-th order operations and H-spaces" by Y. Hemmi to appear in a volume of the series *Contemporary Mathematics*.

Chapter 5

Calculations with Secondary Operations

This chapter develops several examples. The calculations make use of the Milnor filtration and achieve results that are more subtle than those of the previous chapter. In this chapter and subsequently, we write Steenrod operations with parentheses, Sq^a as $Sq(a)$.

5.1. Operations based on $Sq(2)Sq(n)$

We have the Adem relation

$$Sq(2)Sq(n) = \binom{n-1}{2} Sq(n+2) + Sq(n+1)Sq(1),\ n \geq 2$$

with

$$\binom{n-1}{2} = \begin{Bmatrix} 0 & n \equiv 1, 2 \bmod 4 \\ 1 & n \equiv 0, 3 \bmod 4 \end{Bmatrix}.$$

When $n \equiv 1, 2 \bmod 4$, we can construct a stable operation. For the other congruence classes of n, if we leave $Sq(n+2)$ in unfactored form, then we can construct operations that are defined on suitable cohomology classes of dimension less than $(n+2)$. Furthermore, for cohomology classes that are mod 2 reductions of integral classes (even mod 4 classes) the term involving $Sq(1)$ is not needed for the relation. Thus, several possible operations are available and operations can be tailored to the situation at hand. The examples in this section will illustrate nuances of homotopy theory that are captured by tailoring operations to the specific features of the spaces under study.

We start in a context that leads to many informative calculations. Fix the integer n. We shall study operations defined on classes which are mod 2 reductions of integral classes, which are themselves of dimension n, and have cup square 0. In our notation of Chapter 4, we are dealing with operations based on

$$K(Z,n) \xrightarrow{Sq(n)} K(Z/2,2n) \xrightarrow{Sq(2)} K(Z/2,2n+2) .$$

We write the source and target in the usual notation of cohomology theory. Thus

$$\begin{aligned} S_n(X) &= \{x \in H^n(X;Z/2) \mid x \text{ is the reduction of an integral class} \\ &\qquad \text{and } x^2 = 0\} , \\ T_{2n+1}(X) &= H^{2n+1}(X;Z/2)/Sq(2)H^{2n-1}(X;Z/2) . \end{aligned}$$

The set of tetherings is determined by

$$\begin{aligned} H^{2n+1}(K(Z,n);Z/2) =&\{Sq(I)u_n \mid I \text{ admissible not terminating with } 1, \\ &\deg I = n+1, ex(I) < n \text{ and } u_n \text{ is the mod 2 reduction of the} \\ &\text{fundamental class}\} . \end{aligned}$$

We note the absence of product terms.

Several applications will be made to questions of splittings, up to homotopy, for certain suspensions. To prepare for this, we observe that any $(n-1)$-dimensional class

$$y \text{ in } H^{n-1}(Y;Z/2)$$

which is the reduction of an integral class, satisfies

$$\Delta^* y \text{ is in } S_n(\Sigma Y),$$

where Δ^* is the Mayer-Vietoris suspension map. Indeed,

$$\Sigma K(Z,n-1) \xrightarrow{j_{1,\infty}} K(Z,n)$$

is the universal example for such y, by Prop. 1.3.1 and Prop. 3.5.1(d).

Proposition 5.1.1. There exists a secondary operation

$$\Theta_n : S_n(\quad) \to T_{2n+1}(\quad)$$

such that

$$\Theta_n(\Delta^* y) = [\![\Delta^*(y \cup Sq(2)y)]\!]$$

where y is any $(n-1)$-dimensional, mod 2 reduction, of an integral class.

Proof. It is enough to prove the statement for the universal example. We have the following diagram,

$$\begin{array}{ccccccc} \Sigma K(Z,n-1) & \xrightarrow{j_{1,1}} & B_2 & \longrightarrow & (\Sigma K(Z,n-1))^{(2)} & \stackrel{\Sigma p_1}{\to} & \Sigma^2 K(Z,n-1) \\ & & {\scriptstyle j_{2,\infty}}\downarrow & & \downarrow\zeta & & \\ & & K(Z,n) & \underset{Sq(n)}{\longrightarrow} & K(Z/2,2n) & \underset{Sq(2)}{\longrightarrow} & K(Z/2,2n+2) \end{array}$$

where

$$\zeta = [u_{n-1} \mid u_{n-1}]$$

and u_{n-1} is the mod 2 reduction of the fundamental class for $K(Z,n-1)$. The top row comes from the Milnor filtration on $K(Z,n)$ and the square is Prop. 3.5.3(a).

Let H be a tethering for $Sq(2)Sq(n)$ and let

$$\widetilde{Sq(2)} : W_{Sq(n)} \longrightarrow K(Z/2,2n+1)$$

be the colifting constructed from H. By Prop. 3.5.2, there is a lifting

$$\overline{j_{1,\infty}} : \Sigma K(Z,n-1) \longrightarrow W_{Sq(n)}$$

of the composite

$$j_{1,\infty} = j_{2,\infty} \circ j_{1,1}$$

such that the adjoint of

$$z \doteq \widetilde{Sq(2)} \circ \overline{j_{1,\infty}}$$

fits into the square

$$\begin{array}{ccc} (\Sigma K(Z,n-1))^{(2)} & \xrightarrow{\Sigma p_1} & \Sigma^2 K(Z,n-1) \\ \zeta\Big\downarrow & & \Big\downarrow z^\flat \\ K(Z/2,2n) & \underset{Sq(2)}{\longrightarrow} & K(Z/2,2n+2) \end{array}$$

where Σp_1 is from the Hopf construction. Now

$$Sq(2)\zeta = (\Sigma p_1)^*(\Delta_2^*(u_{n-1} \cup Sq(2)u_{n-1}))$$

where the subscript on Δ^* indicates two iterates of this map. Hence

$$z^\flat = \Delta_2^*(u_{n-1} \cup Sq(2)u_{n-1}) + \text{ terms from } \ker(\Sigma p_1)^* .$$

Next, we show that the tethering H can be chosen to eliminate the ambiguous terms. For this, we observe that

$$\ker(\Sigma p_1)^* = \Delta_2^* PH^{2n}(K(Z,n-1);Z/2)$$

where P denotes the submodule of primitive elements. Furthermore, the cohomology suspension

$$\sigma^* : QH^{2n+1}(K(Z,n);Z/2) \to PH^{2n}(K(Z,n-1);Z/2)$$

is an epimorphism. Hence, using (3.3.1) and (1.3.7), we can alter the tethering by elements from

$$PH^{2n+1}(K(Z,n);Z/2)$$

to cancel terms in $\ker(\Sigma p_1)^*$ and add no new ones. This completes the proof of (1.1). □

Our first application of (1.1) is to the question raised in exercise (1.2.1). We use Θ_3 to prove

Proposition 5.1.2. The suspension of CP^3 is not homotopy equivalent to a bouquet $S^7 \vee \Sigma CP^2$.

Proof. Let y generate

$$H^2(CP^3;Z/2)\ .$$

If such an equivalence exists, then there is a self-map f of ΣCP^3 given by the composition

$$\Sigma CP^3 \simeq S^7 \vee \Sigma CP^2 \underset{\text{ret.}}{\longrightarrow} \Sigma CP^2 \underset{\text{inc.}}{\longrightarrow} \Sigma CP^3$$

and satisfying

$$f^*(\Delta^* y) = \Delta^* y,\ f^*(\Delta^* y^3) = 0\ .$$

Since $y^3 = y \cup Sq(2)y$, we obtain a contradiction to (1.1),

$$\begin{aligned} 0 \neq [\![\Delta^* y^3]\!] &= \Theta_3(\Delta^* y) \\ &= \Theta_3(f^*\Delta^* y) \\ &= f^*\Theta_3(\Delta^* y) \\ &= [\![f^*\Delta^* y^3]\!] \\ &= 0\ . \end{aligned}$$

□

Next, we look at loopings of the operation in (1.1).

Proposition 5.1.3. Let y be the mod 2 reduction of an integral class.

(a): If $\dim y = n-1$, then $\Omega\Theta_n$ is defined on y and

$$\Omega\Theta_n(y) = [\![y \cup Sq(2)y]\!] \text{ in } H^{2n}(Y;Z/2)\Big/ imSq(2)\ .$$

(b): If $\dim y = n-2$, then $\Omega^2\Theta_n(y) = 0$, and likewise if $\dim y \leq n-2$.

Proof. Both statements follow from (1.1) and the equation

$$\Delta_k^*\Omega^k\Theta_n(y) = \Theta_n(\Delta_k^* y) = [\![\Delta^*(\Delta_{k-1}^* y \cup Sq(2)\Delta_{k-1}^* y)]\!]\ .$$

□

Remark. We can apply (1.3) to $SU(3)$ to discover that the secondary operation Θ_4 is non-trivial on $\Sigma SU(3)$. Thus, the top cell for $SU(3)$ does not split off after one suspension. The further investigation of splittings under iterated suspensions of $SU(3)$, or complex projective spaces, will illustrate the advantage to tailoring operations to specific situations.

Now we place the integer n in the congruence class of 1,2 mod (4) and consider the problem of extending the initial segment of operations

$$(\Omega^2\Theta_n, \Omega\Theta_n, \Theta_n)$$

to a stable operation. The situation here differs from the general construction in Prop. 4.2.8, because we do not wish to receive Θ_n as the result of looping down, but insist that, as a component of a stable operation, it be the operation of (1.1). Of course, it can transpire, that it does not matter which way we go, but in any case, analysis is required.

We now set up indexing for the component of a stable operation so that Θ_n will be the component with $m = 0$. Set

$$\begin{aligned} C_{0,m} &= K(Z, n+m)\ , \\ C_{1,m} &= K(Z/2, 2n+m)\ , \\ C_{2,m} &= K(Z/2, 2n+2+m) \end{aligned}$$

with

$$\theta_m : C_{0,m} \to C_{1,m} \text{ given by } Sq(n)$$

and

$$\varphi_m : C_{1,m} \to C_{2,m} \text{ given by } Sq(2)$$

where $n \equiv 1, 2 \bmod 4$. We write

$$\Theta_{n,m} : S_{\theta_m}(\quad) \to T_{\Omega\varphi_m}(\quad)$$

to signify an operation based on $[\varphi_m][\theta_m] = 0$. For $m = -2, -1, 0$, we assume $\Theta_{n,m}$ is the operation of (1.1) and its loops. We omit τ_m since θ_m and $-\theta_m$ are homotopic.

For the first step in the extension of our initial segment, we compare

$$\Theta_{n,0} \text{ with } \Omega\Theta_{n,1}\ .$$

The difference is measured by an element in

$$[C_{0,0}, \Omega C_{2,0}] = H^{2n+1}(K(Z,n); Z/2)\ .$$

We have described this set and noted its absence of products. In particular, every element is a loop class

$$\Omega\delta \text{ for some } \delta : C_{0,1} \to \Omega C_{2,1}\ .$$

Thus, if necessary, the tethering for $\Theta_{n,1}$ may be altered so that

$$\Omega\Theta_{n,1} = \Theta_{n,0}$$

holds. Moreover, the condition that $\Theta_{n,1}$ be compatible with exact sequences holds, because $\Theta_{n,1}$ is associated to a sequence with homotopy.

Next, we compare

$$\Omega\Theta_{n,2} \text{ with } \Theta_{n,1} \ .$$

The difference is measured by

$$[C_{0,1}, \Omega C_{2,1}] = H^{2n+2}(K(Z, n+1); Z/2) \ .$$

Again each such element is a loop class and, if necessary, the tethering for $\Theta_{n,2}$ may be altered so that

$$\Omega\Theta_{n,2} = \Theta_{n,1} \ .$$

The stability hypothesis (4.2.7) holds for $m \geq 2$, so we can use the general principles of Prop. 4.2.8 to extend the segment constructed above

$$(\Omega^2\Theta_n, \ \Omega\Theta_n, \ \Theta_n, \ \Theta_{n,1}, \ \Theta_{n,2})$$

to a stable operation.

Summarizing this discussion, we have

Proposition 5.1.4. If $n \equiv 1, 2 \bmod (4)$, there is a stable secondary operation

$$\{\Theta_{n,m}\}, m \geq -2$$

based on $Sq(2)Sq(n) = 0 \bmod Sq(1)$ such that the domain of $\Theta_{n,m}$ consists of $(n+m)$-dimensional classes which are the mod 2 reduction of integral classes and are annihilated by $Sq(n)$. The values of $\Theta_{n,m}$ lie in the quotient

$$H^{2n+1+m}(X; Z/2) \Big/ im\, Sq(2)$$

and for $m = 0$, $\Theta_{n,0}$ is the operation Θ_n described in (1.1).

Remark. The reader will have noticed that the universal example for

$$\Theta_{n,-1} = \Omega\Theta_{n,0}$$

has the homotopy type of

$$K(Z, n-1) \times K(Z/2, 2n-2) \ .$$

The non-trivial value of $\Theta_{n,-1}$ on the mod 2 reduction of the fundamental class, u_{n-1}, indicates that this splitting is not compatible with the loop structure. We shall return to this topic in Chapter 7, where it will enter into the discussion of the cohomology structure of universal examples. Moreover, tetherings for $\Theta_{n,-1}$ are restricted to image Ω.

We turn now to the cases where n is congruent to 0,3 mod (4). Our original Adem relation has an unfactored term

$$Sq(n+2)\ .$$

Of course, this term is always factorable, but we must consider whether there is a factorization which is different from the relation in use. The cases $n \equiv 0(4)$ and $n \equiv 3(4)$ differ in this respect. For $n \equiv 3(4)$, we have the Adem relation

$$Sq(1)Sq(n+1) = Sq(n+2)\ .$$

Hence we have

$$Sq(2)Sq(n) + Sq(1)Sq(n+1) \equiv 0 \text{ mod } (Sq(1))\ .$$

For $n \equiv 0(4)$, e.g. $n = 4$, attempts to rewrite $Sq(n+2)$ will affect the term with the $Sq(2)$ factor, e.g.

$$Sq(6) = Sq(2)Sq(4) + Sq(5)Sq(1)$$

and this is the only relation up to factoring $Sq(5)$. It will turn out, that this difference in algebra is supported by differences in homotopy theory.

We turn to the analogue of Prop. 5.1.4 for the case $n \equiv 3(4)$. Write

$$\begin{aligned}
C_{0,m} &= K(Z, n+m)\ ,\\
C_{1,m} &= K(Z/2, 2n+m) \times K(Z/2, 2n+1+m)\ ,\\
C_{2,m} &= K(Z/2, 2n+2+m)\ ,\\
\theta_m : C_{0,m} &\to C_{1,m} \text{ given by } \begin{pmatrix} Sq(n) \\ Sq(n+1) \end{pmatrix}\ ,\\
\varphi_m : C_{1,m} &\to C_{2,m} \text{ given by } (Sq(2), Sq(1))\ .
\end{aligned}$$

Proposition 5.1.5. There is a stable operation $\{\Phi_{n,m}\}$,

$$\Phi_{n,m} : S_{\theta_m}(\quad) \to T_{\Omega\varphi_m}(\quad) \text{ for } n \equiv 3(4)\ .$$

On $(n-1)$-dimensional, mod 2 reduction of integral classes, y we have

$$\Phi_{n,0}(\Delta^* y) = [\![\Delta^*(y \cup Sq(2)y)]\!]\ .$$

Proof. The argument is a reprise of the steps in this section. One begins by observing that the change of the space $C_{1,0}$ does not affect the proof of (1.1) because

$$Sq(n+1)u_n = 0\ .$$

We leave the remaining verifications for an exercise. □

We can refine Prop. 5.1.2 with the following.

Exercise 5.1.6. No suspension of CP^3 splits non-trivially.

Exercise 5.1.7. No suspension of CP^k splits non-trivially. (Hint): The key is to combine the action of $Sq(2)$ and $Sq(4)$ with

$$\Omega\Phi_{4t+3,0},\ t \geq 0\,,$$

on y^{2t+1}, where y generates $H^2(CP^k; Z/2)$. Since

$$Sq(2)y^{2t+1} = y^{2t+2}$$

we have

$$\Omega\Phi_{4t+3,0}(y^{2t+1}) = [\![y^{4t+3}]\!] \neq 0\,.$$

For $n \equiv 0(4)$, we can make the first step in the argument for Prop. 5.1.4. Thus we have

$$\Omega\Theta_{n,1} = \Theta_{n,0}$$

and

$$\Theta_{n,1}(\Delta_2^* y) = [\![\Delta_2^*(y \cup Sq(2)y)]\!]$$

where y is of dimension $(n-1)$ and the mod 2 reduction of an integral class.

However, it is a fact that

$$\Sigma^3 SU(3) \simeq \Sigma^4 CP^2 \vee S^{11},$$

so no further delooping of Θ_4 is possible. For larger values of n (still $\equiv 0(4)$), the impossibility of a stable operation is revealed by

$$SU(2m+1)/SU(2m-1)\,.$$

The mod 2 cohomology is an exterior algebra

$$\Lambda(y_{4m-1},\ Sq(2)y_{4m-1})$$

and the top cell is stably spherical.

Remark. Application of the operations discussed above to immersion questions for manifolds is given in [**68**]. One of the main calculations in [**68**] is treated in subsection (7.3.2).

Our study has carried a hypothesis which serves to eliminate the need to include the term

$$Sq(n+1)Sq(1)$$

in the defining relation for a secondary operation. This is convenient, and we ask whether it is necessary. The answer is that it is necessary. By way of explanation, we can view Prop. 5.1.1 and its extensions as a decomposition of a certain primary operation by a secondary operation. The possibility for such a decomposition changes when the $Sq(n+1)Sq(1)$ term is present. In more detail, consider the operation

$$y \to [\![y \cup Sq(2)]\!]$$

where y is a mod 2 cohomology class and the cup product is projected into a quotient. In this case where y is the mod 2 reduction of an integral class, we have expressed

$$y \to [\![y \cup Sq(2)]\!] \text{ in } H^*(Y)/im\,Sq(2)$$

in terms of a component of a stable operation in the cases where

$$\dim y \equiv 0, 1, 2 \bmod 4$$

and observed that such an expression is impossible in the remaining congruence class 3(4). In fact, this impossibility is of homotopy theoretic origin. In [**71**] it is proved, that for sufficiently many suspensions, there is a non-trivial composition

$$S^{8n+8+k} \to \Sigma^k K(Z/2, 4n+3) \to K(Z/2, 8n+8+k)$$

where the second map represents the k-fold suspension of

$$u_{4n+3} \cup Sq(2)u_{4n+3}\,.$$

The situation changes for $K(Z/2, n)$. It turns out that there is a non-trivial composition

$$S^{2n+2+k} \to \Sigma^k K(Z/2, n) \to K(Z/2, 2n+2+k)$$

detecting

$$\Delta_k^*(b_n \cup Sq(2)b_n)$$

for k sufficiently large and $n \equiv 0, 2, 3(4)$, but not for $n \equiv 1(4)$. This result appears in an unpublished portion of the author's thesis (University of Chicago, 1967), directed by A.L. Liulevicius).

For later reference, here is the data for the operation involved where $n \equiv 1(4)$. We make full use of the notation introduced in subsection (1.3.6). We have

$$\Sigma^{m-n} K(Z/2, n) \xrightarrow[\epsilon]{} K(m, m+n+1) \xrightarrow[\theta_m]{} K(m+n+1, m+n+2) \xrightarrow[\varphi_m]{} K(m+n+3)$$

where

$$\epsilon = \begin{pmatrix} \Delta_{m-n}^* & b_n \\ \Delta_{m-n}^* & (b_n \cup Sq(1)b_n) \end{pmatrix},$$

$$\theta_m = \begin{pmatrix} Sq(n+1), & 0 \\ Sq(n+1,1), & Sq(1) \end{pmatrix},$$

and

$$\varphi_m = (Sq(2), Sq(1))\,.$$

Then there is a stable secondary operation

$$\Lambda_m : S_{\theta_m}(\quad) \to T_{\Omega\varphi_m}(\quad), m \geq n$$

and

$$\Lambda_m(\epsilon) = [\![\Delta_{m-n}^*(b_n \cup Sq(2)b_n)]\!] \neq 0\,.$$

Exercise 5.1.8. Investigate the operation based on

(a) $$Sq(1)Sq(n) = \binom{n-1}{1} Sq(n+1) .$$

Show there is a choice Ψ_n such that for any mod 2 class y of dimension $(n-1)$, we have

$$\Psi_n(\Delta^* y) = [\![\Delta^*(y \cup Sq(1)y)]\!] .$$

Use this fact to prove that

$$\Sigma RP^3 \not\simeq \Sigma RP^2 \vee S^4 .$$

Likewise, investigate the operation based on

(b) $$Sq(2)Sq(n) + Sq(n+1)Sq(1) = \binom{n-1}{2} Sq(n+2) .$$

Show there is a choice Φ_n satisfying

$$\Phi_n(\Delta^* u_n) = [\![\Delta^*(u_n \cup Sq(2)u_n)]\!]$$

where

$$u_n \text{ in } H^n(K(Z/4, n); Z/2)$$

is the mod 2 reduction of the fundamental class.

Remark. The work of this section is guided by a single strategy – compatibility with exact sequences for the map obtained from the Hopf construction on loop multiplication

$$\Omega X \times \Omega X \to \Omega X$$

and the observation that our operations are to be evaluated on a map

$$\epsilon : \Sigma\Omega X \to C_0$$

which factors through $B_2 X$. The strategy fails for the operation Λ_m described above because its ϵ does not factor this way. The stated result is proved using methods developed in Chapter 7. The effectiveness of those methods lies in placing elements constructed geometrically in the cohomology of the universal example.

5.2. Higher order Bocksteins

5.2.1. Higher order Bocksteins and p-th powers. Let f be an integer, $f \geq 2$. We discuss the representation of a generator for

$$H^{n+1}(K(Z/p^f, n);\ Z/p) = Z/p,\ p \text{ a prime},$$

by means of higher order operations. The construction of Bocksteins, as coefficient connecting homomorphisms, using the diagram

$$\begin{array}{ccccccccc} 0 & \longrightarrow & Z/p & \longrightarrow & Z/p^{f+1} & \longrightarrow & Z/p^f & \longrightarrow & 0 \\ & & \downarrow j & & \| & & \downarrow \rho & & \\ 0 & \longrightarrow & Z/p^f & \longrightarrow & Z/p^{f+1} & \longrightarrow & Z/p & \longrightarrow & 0 \end{array},$$

and regarding the results as maps of Eilenberg-Mac Lane spaces, yields a homotopy commutative diagram,

$$\begin{array}{ccc} K(Z/p^f, n) & \xrightarrow{\beta_f} & K(Z/p, n+1) \\ \rho_\# \downarrow & & \downarrow j_\# \\ K(Z/p, n) & \xrightarrow[\delta_f]{} & K(Z/p^f, n+1) \end{array},$$

where the subscript on ρ, j indicates an induced coefficient homomorphism.

In particular, in terms of cochains, β_f is defined as follows. Let χ be a singular Z/p^f-cocycle representing a Z/p^f-cohomology class, and lift χ to a singular Z/p^{f+1}-cochain, $\bar{\chi}$. The coboundary of $\bar{\chi}$ factors through Z/p, yielding a Z/p-cocycle ξ, as in the diagram

$$\begin{array}{ccccc} S_{n+1}(X) & \xrightarrow{\partial} & S_n(X) & = & S_n(X) \\ \xi \downarrow & & \downarrow \bar{\chi} & & \downarrow \chi \\ Z/p & \longrightarrow & Z/p^{f+1} & \longrightarrow & Z/p^f \end{array}.$$

Then ξ represents $\beta_f([\chi])$. From the definition, we have the following facts.

(a): For y in $H^*(Y; Z/p^f)$, then py is in the image of $j_\#$ for

$$j : Z/p^{f-1} \to Z/p^f$$

and

$$\beta_f(py) = \beta_{f-1}(y_1)$$

where $j_\# y_1 = py$. Furthermore, the equation is independent of the choice of y_1.

(b): For y, y' in $H^*(Y; Z/p^f)$, there is a Cartan formula

$$\beta_f(yy') = \beta_f(y) \cup \rho_\# y' + (-1)^{|y|} \rho_\# y \cup \beta_f(y')$$

where $|y|$ denotes the degree of y. There is no indeterminancy in this formula, because the Bocksteins are calculated on Z/p^f-classes.

In principle, we may calculate β_f either in terms of maps, or through the cochain definition. We write this fact as

$$\beta_f(\epsilon) = [\![\beta_f(y)]\!] \text{ in } T_{\beta_{f-1}}(Y)$$

where $\rho_{\#}y$ is represented by

$$\epsilon : Y \to K(Z/p, n) .$$

Mixing notation, we write (b) as

$$\beta_f(\rho_{\#}yy') = [\![\beta_f(\rho_{\#}y) \cup \rho_{\#}y' + (-1)^{|y|}\rho_{\#}y \cup \beta_f(\rho_{\#}y')]\!] .$$

Our goal is to calculate β_f on the p-th power u_{2n}^p where

$$u_{2n} \text{ in } H^{2n}(K(Z_{p^{f-1}}, 2n);\ Z/p)$$

is the mod p reduction of the fundamental class. This calculation was first produced by Browder [**13**], making use of Cartan's calculations [**20**].

We have β_f displayed as a higher order operation

$$\beta_f : S_{\rho_{\#}}(\quad) \to T_{\beta_{f-1}}(\quad)$$

with a buttrussing supplied by δ_f. In order to make Browder's calculation, we develop a description of β_f as a higher order operation associated with a syzygy of length f. Our discussion will use

(a) $H^{n+1}(K(Z/p^f, n); Z/p) = Z/p$,

$H^{n+1}(K(Z/p, n); Z/p^f) = Z/p$ for $n \geq 2$,

(b) $$H^{n+2}(K(Z/p^f, n); Z/p) = \begin{Bmatrix} 0 & p \text{ odd} \\ \text{Span } \{Sq(2)u_n\} & p = 2 \end{Bmatrix} .$$

Proof. We observe that for $f \geq 1$,

$$H_{n+1}(K(Z/p^f, n); Z) = 0 .$$

For $n \geq 2$, this fact follows because the Hurewicz map from π_{n+1} to H_{n+1} is surjective. For $n = 1$, we may use Hopf's formula for the second homology. Then (a) follows from the Universal Coefficient Theorem. Part (b) follows from the Whitehead sequence

$$\cdots \to \pi_{n+2} \to H_{n+2} \to \Gamma_{n+1} \to \pi_{n+1} \to \cdots$$

where

$$\Gamma_{n+1} = im\{\pi_{n+1}X^n \to \pi_{n+1}X^{n+1}\}$$

and $K(Z/p^f, n)$ has the homotopy type of a CW complex

$$X = S^n \cup_{p^f} e^{n+1} \cup \{\text{cells in dimension } \geq n+2\}.$$

In the sequel, we take $n \geq 2$. From (b), it follows that any sequence of Bocksteins (first order after the initial map) gives a syzygy,

$$K(Z/p^k, n) \underset{\beta_k}{\longrightarrow} K(Z/p, n{+}1) \underset{\beta}{\longrightarrow} K(Z/p, n{+}2) \underset{\beta}{\longrightarrow} K(Z/p, n{+}3) \longrightarrow \cdots \ .$$

Note that for $p = 2$, $Sq(1)\beta_k = 0$ because $Sq(2)$ is non-zero on

$$\Sigma^{p-2} CP^2 \ .$$

Differences between contracting homotopies for the composition

$$\beta \circ \beta_k$$

factor through β_k. Hence, up to homotopy, there is a unique colifting

$$\tilde{\beta} : W_{\beta_k} = K(Z/p^{k+1}, n) \longrightarrow K(Z/p, n+1) \ .$$

The colifting $\tilde{\beta}$ is non-trivial, because $\Omega\beta$ is non-trivial. Furthermore, from (a), any colifting agrees, up to non-zero constant in Z/p, with the coefficient connecting homomorphism. Moreover, from (b) we obtain the information that the syzygy is admissible because

$$\Omega\beta \circ \tilde{\beta} : K(Z/p^{k+1}, n) \longrightarrow K(Z/p, n+2)$$

is null homotopic.

We also wish to have stability, in the sense that

$$\Omega\beta_f = \beta_f$$

as n is varied. Thus with n fixed and β_{f-1} assumed by induction, we define β_f as the colifting based on

$$(\boldsymbol{\beta}, \tau_n \circ \beta_{f-1})$$

where $\boldsymbol{\beta}$ is the signed Bockstein and τ_n is defined in (4.2.6). We then choose δ_f to buttruss, β_f, as (a) guarantees that such a choice is possible. Then the stability property for β_f entails the same for δ_f,

$$\Omega\delta_f = \delta_f \ .$$

The following diagram displays our constructions,

$$\begin{array}{ccccc}
K(Z/p^f, n) & \xrightarrow{\beta_f} & K(Z/p, n+1) & & \\
\downarrow & & & & \\
K(Z/p^{f-1}, n) & \xrightarrow{\tau_n \circ \beta_{f-1}} & K(Z/p, n+1) & \xrightarrow{\boldsymbol{\beta}} & K(Z/p, n+2) \\
\downarrow & & \downarrow & & \downarrow \lambda_\# \\
K(Z/p, n) & \underset{\tau_n \circ \delta_{f-1}}{\longrightarrow} & K(Z/p^{f-1}, n+1) & \underset{\beta_{f-1}}{\longrightarrow} & K(Z/p, n+2)
\end{array}$$

where $\lambda_{\#}$ is induced by multiplication by some non-zero element in Z/p. Thus, given

$$\epsilon : Y \to K(Z/p, n)$$

which lifts to

$$\bar{\epsilon} : Y \to K(Z/p^f, n)$$

we have

$$\beta_f(\epsilon) = [\![\beta_f \circ \bar{\epsilon}]\!] \text{ in } T_{\beta_{f-1}}(Y) .$$

□

Theorem 5.2.1 (Browder's Theorem). **(a):** *For $p = 2$ and $f = 2$, the secondary Bockstein is defined on b_{2n}^2, where b_{2n} is the fundamental class of $K(Z/2, 2n)$ and*

$$\beta_2(b_{2n}^2) = [\![b_{2n} \cup Sq(1)b_{2n} + Sq(2n, 1)b_{2n}]\!]$$

in

$$H^{4n+1}(K(Z/2, 2n);\ Z/2)\Big/ imSq(1) .$$

(b): *For $f \geq 2$ with p odd or $f \geq 3$ with $p = 2$, the higher Bockstein β_f is defined on u_{2n}^p, where u_{2n} is the mod p reduction of the fundamental class for $K(Z/f^{f-1}, 2n)$ and*

$$\beta_f(u_{2n}^p) = \lambda[\![u_{2n}^{p-1} \cup \beta_{f-1}u_{2n}]\!]$$

in

$$H^{2np+1}(K(Z/p^{f-1}, 2n); Z/p)/im\beta_{f-1}$$

and λ is a non-zero element of Z/p.

Proof. We can write

$$u_{2n}^p = \mathcal{P}(n)u_{2n} \text{ or } b_{2n}^2 = Sq(2n)b_{2n} .$$

Thus we wish to calculate

$$\beta_f\mathcal{P}(n)u_{2n}$$

when it is defined. First we treat the case $f = 2$, and p any prime. We use the information in Prop. 3.5.3 for $K(Z/p, 2n+1)$. Thus we have the following diagram, commuting up to homotopy,

$$\begin{array}{ccccccccc}
 & & \Sigma K(Z/p,2n) & \xrightarrow{j_{1,1}} & B_2 & \longrightarrow & (\Sigma K(Z/p,2n))^{(2)} & \xrightarrow{\Sigma p_1} & \Sigma^2 K(Z/p,2n) \\
 & & \| & & \downarrow{\scriptstyle j_{2,\infty}} & & \downarrow{\scriptstyle \zeta} & & \\
 & & \Sigma K(Z/p,2n) & \xrightarrow{j_{1,\infty}} & K(Z/p,2n+1) & \xrightarrow[\tau\circ\beta\circ\mathcal{P}(n)]{} & K(Z/p,2np+2) & & \\
 & & & & \downarrow{\scriptstyle \mathcal{P}(n)} & & \| & & \\
K(Z/p^2,2np+1) & \to & & & K(Z/p,2np+1) & \xrightarrow[\tau\circ\beta]{} & K(Z/p,2np+2) & \xrightarrow[\beta]{} & K(Z/p,2np+3) \\
\downarrow{\scriptstyle \beta_2} & & & & & & & & \\
K(Z/p,2np+2) & . & & & & & & &
\end{array}$$

The value of ζ is

$$\zeta = \lambda \sum_{t=1}^{p-1} \frac{1}{p} \binom{p}{t} [b_{2n}^t \mid b_{2n}^{p-t}] \qquad \text{for } p \text{ odd, } \lambda \neq 0 \,,$$

$$\zeta = [b_{2n} \mid b_{2n}] \qquad p = 2 \,.$$

The theory in Prop. 3.5.2 provides a lifting of $\mathcal{P}(n) \circ j_{1,\infty}$ with the property that the adjoint of its composition with β_2 satisfies

$$\left(\beta_2 \circ \overline{\mathcal{P}(n) \circ j_{1,\infty}}\right)^\flat \circ \sum p_1 = (-1) \text{ times } \beta \circ \zeta \,.$$

(Here, we know $\lambda = -1$, but its precise value at later stages is fudged in our buttruss $\lambda_{\#}$.) Now writing in terms of cohomology, a direct calculation using the Cartan formula for β, yields

$$\beta \circ \zeta = (\Sigma p_1)^* (\Delta_2^* (b_{2n}^{p-1} \cup \beta b_{2n} + \text{ prim.}))$$

where Δ_2^* is two iterates of the Mayer-Vietoris suspension and the unknown term written "prim." is an element of the submodule

$$PH^{2np+1}(K(Z/p, 2n); Z/p) \,.$$

To find the value of this extra term, we use the information that composing β_2 with β yields 0. Hence

$$0 = \beta(\text{prim.}) \text{ for } p \text{ odd}$$

and

$$0 = (Sq(1) b_{2n})^2 + Sq(1) \text{ prim.} \qquad \text{for } p = 2.$$

Hence for $p = 2$,

$$\text{prim.} \ = Sq(2n, 1) b_{2n} + Sq(1) x_1$$

and for p odd

$$\text{prim.} \ = \beta x_2$$

using (1.5.3)(d), from the sundry facts, and part (a) is established.

For the cases $f \geq 3$, we argue by induction on f, assuming the result for $k \leq f - 1$. Thus

$$\beta_k(\mathcal{P}(n) u_{2n}) = \lambda [\![u_{2n}^{p-1} \cup \beta_{k-1} u_{2n}]\!] = 0$$

and likewise for $p = 2$. Hence, we may choose a lifting of

$$\mathcal{P}(n) u_{2n}$$

which is in

$$S_{\beta_{f-1}}(K(Z/p^{f-1}, 2np))$$

for

$$\beta_{f-1} : K(Z/p^{f-1}, 2np) \to K(Z/p, 2np + 1) \,.$$

Since β_{f-1} is buttrussed by δ_{f-1}, it follows that

$$\delta_{f-1} \circ \mathcal{P}(n) = 0$$

for

$$\mathcal{P}(n) : K(Z/p^{f-1}, 2n) \to K(Z/p, 2np) .$$

We do not develop a result like those in subsection (3.5.6) for this fact, but use it in a different way. To set things up, we abbreviate notation to write

$$K(k, m) \text{ for } K(Z/p^k, m) .$$

We study a tower that is one delooping back from where we wish to calculate,

$$\begin{array}{ccccccc} K(f-1, 2np+1) & \xrightarrow{j_1} & E_{f-1} & \xrightarrow{j_2} & K(f, 2np+1) & \xrightarrow{\beta_f} & K(1, 2np+2) \\ & & \downarrow{\scriptstyle q} & & \downarrow{\scriptstyle \rho\#} & & \\ & & K(f-1, 2n+1) & \xrightarrow[\mathcal{P}(n)]{} & K(1, 2np+1) & \xrightarrow[\delta_{f-1}]{} & K(f-1, 2np+2) . \end{array}$$

The square is a pull-back and the composition across the top is β_{f-1}. By naturality of pull-backs, we may regard the space E_{f-1} as the homotopy fiber of the composition

$$\delta_{f-1} \circ \mathcal{P}(n).$$

Thus, after looping, we have the diagram

$$\begin{array}{ccccc} K(f-1, 2np) & \xrightarrow{\Omega j_1} & \Omega E_{f-1} & \xrightarrow{\tilde{I}} & K(f-1, 2np) \\ & & \Big\downarrow{\scriptstyle \Omega q} & & \\ & & K(f-1, 2n) & & \end{array}$$

with $\tilde{I} \circ \Omega j_1 \simeq$ identity map. We can add these maps to obtain a homotopy equivalence (of spaces, not necessarily H-spaces)

$$\Omega E_{f-1} \xrightarrow{\Delta} \Omega E_{f-1} \times \Omega E_{f-1} \xrightarrow{\Omega q \times \tilde{I}} K(f-1, 2n) \times K(f-1, 2np) .$$

We shall make our calculations in terms of this splitting. In terms of cohomology, $\beta_f \circ \Omega j_2$ represents a primitive cohomology class

$$v'_{2np+1} \text{ in } H^{2np+1}(\Omega E_{f-1}; Z/p)$$

such that

$$(\Omega j_1)^* v'_{2np+1} \text{ represents } \beta_{f-1} u_{2np}$$
$$\text{in } H^{2np+1}(K(f-1, 2np); Z/p) .$$

Let $s : K(f-1, 2n) \to \Omega E_{f-1}$ be a section for Ωq. Then

$$\beta_f(u_{2n}^p) = [\![s^* v'_{2np+1}]\!] \text{ in } T_{\beta_{f-1}}(K(f-1, 2n)) .$$

Let u'_{2np} denote the mod p reduction of the fundamental class of $K(f-1, 2np)$ in the splitting of ΩE_{f-1}. Then

$$(\Omega j_1)^* u'_{2np} = u_{2np} \, .$$

To determine the primitive class v'_{2np+1}, we require the coproduct of $u'_{2r\text{-}p}$ induced by loop multiplication. The answer is

$$u'_{2np} \to 1 \otimes u'_{2np} + u'_{2np} \otimes 1 + \lambda \sum_{t=1}^{p-1} \frac{1}{p} \binom{p}{t} u^t_{2n} \otimes u^{p-t}_{2n}, \ \lambda \neq 0 \, .$$

Sufficient information for this calculation is available in Prop. 3.5.3(b) for $K(f-1, 2n+1)$ combined with the Milnor filtration on E_{f-1}. We have the following diagram,

$$\begin{array}{ccccc}
B_2 E_{f-1} & \longrightarrow & (\Sigma \Omega E_{f-1})^{(2)} & \xrightarrow{\Sigma p_1} & \Sigma^2 \Omega E_{f-1} \\
\downarrow & & \downarrow & & \\
B_2 & \longrightarrow & (\Sigma K(f-1, 2n))^{(2)} & & \downarrow \zeta' \\
\downarrow & & \downarrow \zeta & & \\
K(f-1, 2n+1) & \xrightarrow[-\beta \mathcal{P}(n)]{} & K(1, 2np+2) & = & K(1, 2np+2)
\end{array}$$

where, by (3.5.3b)

$$\zeta = -\sum_{t=1}^{p-1} \frac{1}{p} \binom{p}{t} [u^t_{2n} \mid u^{p-t}_{2n}] \, .$$

The double adjoint of ζ' has the coproduct determined by ζ. Since no term from the mod p cohomology of $K(f-1, 2n)$ can produce a coproduct of this form, we have

$$(\zeta')^\flat = \lambda u'_{2np} + \text{ terms from } PH^{2np}(K(f-1, 2n); Z/p), \lambda \neq 0 \, .$$

Now, using the Cartan formula for β_{f-1}, we have

$$[\![v'_{2np+1}]\!] = [\![\beta_{f-1} u'_{2np} - \lambda u^{p-1}_{2n} \cup \beta_{f-1} u_{2n} + \text{ prim.}]\!] \, .$$

We trim the ambiguity from unknown primitives by using the fact that

$$\beta v'_{2np+1} = 0 \, .$$

Then part (b) is obtained using (1.5.3)(d) as in the first part of our proof. □

Remark. We did not use material from subsection (3.5.7) in this argument, as it is not necessary. However, the extra information is needed in section 5.

5.2.2. Some examples. The most far reaching consequences of Browder's Theorem are in the study of finite H-spaces. The books by Kane [**48**], Zabrodsky [**125**] and papers by J. Lin [**57**], [**58**] develop this subject. The result proved above is the basis for a theory of "infinite implications" in the classical Bockstein spectral sequence for H-spaces. This development appears in [**13**]. An interesting open question concerning a Bockstein type spectral sequence for a generalized cohomology theory is formulated by Kane in [**50**]. We turn to examples of another kind.

The mod p cohomology of the three-connected cover of S^3 is given by

$$H^*(S^3\langle 3\rangle; Z/p) = Z/p[x_{2p}] \otimes \Lambda(x_{2p+1})$$

with

$$\beta x_{2p} = x_{2p+1} \; .$$

Then we have

$$\begin{aligned} \beta_2 x_{2p}^p &= \lambda x_{2p}^{p-1} \cup x_{2p+1} \; , \\ \beta_3 x_{2p}^{p^2} &= \lambda x_{2p}^{p^2-1} \cup x_{2p+1} \; , \quad \lambda \neq 0 \; . \end{aligned}$$

and so on through the powers of p. These formulas hold with no indeterminacy.

As another example, let X be a space with mod 2 cohomology satisfying the following condition; there is a $2n$-dimensional cohomology class x_{2n} such that

$$x_{2n}^2 = 0 \; .$$

Then, we have the following relation in the mod 2 cohomology of X,

$$x_{2n} \cup Sq(1)x_{2n} + Sq(2n,1)x_{2n} = Sq(1)y$$

for some $4n$-dimensional class y.

This equation holds because

$$0 = \beta_2(x_{2n}^2) \; .$$

This relation rules out finding spaces X with

$$H^*(X; Z/2) \cong \Lambda(x_{2n}, Sq(1)x_{2n})$$

for $n \neq 1$. Since $H^{4n}(X) = 0$, we must have

$$x_{2n} \cup Sq(1)x_{2n} = Sq(2n,1)x_{2n} \; .$$

If $n > 1$, then $Sq(2n,1)x_{2n} = Sq(2)Sq(2n-1)x_{2n} = 0$ because $H^{4n-1}(X) = 0$.

When $n = 1$ and $H^*(X) = \Lambda(x_2, Sq(1)x_2)$, then we also have

$$x_2 \cup Sq(1)x_2 = Sq(2,1)x_2 \; .$$

5.3. The Adams operations

We have observed that the mod 2 Steenrod algebra is generated by $Sq(2^i), i = 0, 1, 2, \ldots$. We study operations based on Adem relations for these generators. The operations which are defined on classes which are annihilated by each $Sq(2^i)$ (and hence all Steenrod operations of positive degree) are known as *Adams operations*. They play a basic role in the decomposition of $Sq(2^N)$, $N \geq 4$, by secondary operations.

In this section, we construct the basic Adams operations

$$\Phi_{i,j},\ 0 \leq i \leq j,\ j \neq i+1$$

and evaluate $\Phi_{0,j}$ on certain classes in the mod 2 cohomology of CP^∞. In the next section, we give a formal discussion of the sense in which these operations are basic. Here the term "basic" has informal status.

We start with an Adem relation

$$Sq(2^b)Sq(2^b) = \sum_{t=0}^{b-1} Sq(2^{b+1} - 2^t)Sq(2^t)\ .$$

Proof. The binomial coefficient satisfies

$$\binom{2^b - 1 - t}{2^b - 2t} = \left\{\begin{matrix} 0 & t \text{ not a power of } 2 \\ 1 & t \text{ a power of } 2 \end{matrix}\right\} \bmod 2\ .$$

□

Since we have used all the relevant generators for the Steenrod algebra, our general theory yields

Proposition 5.3.1. There is a unique stable secondary operation

$$\Phi_{b,b}, \quad b \geq 0$$

based on the relation

$$\sum_{t=0}^{b} Sq(2^{b+1} - 2^t)Sq(2^t) = 0\ .$$

Exercise 5.3.2. Prove that $2\iota, \eta^2, \nu^2, \sigma^2$ are detected by $\Phi_{b,b}$ for $b = 0, 1, 2, 3$ respectively.

Next we aim for the operation $\Phi_{0,b}$. We have the Adem relations

$$Sq(1)Sq(2^b) = Sq(2^b + 1)$$

and

$$Sq(2)Sq(2^b - 1) = Sq(2^b + 1) + Sq(2^b)Sq(1) \text{ if } b \geq 2\ .$$

Hence we have

5.3.1. Relation.

$$Sq(1)Sq(2^b) + Sq(2)Sq(2^b - 1) + Sq(2^b)Sq(1) = 0,\ b \geq 2\ .$$

Secondary operations based on the relation under subheading (5.3.1) are useful in the study of spaces with cohomology concentrated in even dimensions. We shall use it in the development of the Adams decomposition formula. In other applications, we want an operation to be based on a relation where the entries in each doubleton summand are powers of 2. Then our operation will have the same domain as in Prop. 5.3.1. To achieve this, we observe

$$Sq(2^b - 1) = Sq(1)Sq(2)\cdots Sq(2^{b-1}) = Sq(2^{b-1} - 1)Sq(2^{b-1})\ .$$

Hence we can replace $Sq(2)Sq(2^b - 1)$ with

$$Sq(2)Sq(2^{b-1} - 1)Sq(2^{b-1})\ ,$$

and for $b \geq 3$ we have

$$Sq(2)Sq(2^{b-1} - 1) = Sq(1)Sq(2^{b-1}) + Sq(2^{b-1})Sq(1)\ .$$

Proposition 5.3.3. There is a unique stable secondary operation

$$\Phi_{0,b}\ ,\ b \geq 2$$

based on the relation

$$\begin{gathered} Sq(4)Sq(1) + Sq(2,1)Sq(2) + Sq(1)Sq(4) = 0 \text{ for } b = 2\ , \\ Sq(2^b)Sq(1) + [Sq(1)Sq(2^{b-1}) + Sq(2^{b-1})Sq(1)]Sq(2^{b-1}) \\ + Sq(1)Sq(2^b) = 0 \text{ for } b \geq 3\ . \end{gathered}$$

Exercise 5.3.4. Prove that 2ν and 2σ are detected by $\Phi_{0,b}$ for $b = 2, 3$ respectively.

We turn to the remaining Adams operations. If $a < b$, we have the Adem relation

$$Sq(2^a)Sq(2^b) = Sq(2^a + 2^b) + \sum_{2^{a-1} \geq t \geq 1} \lambda_t Sq(2^a + 2^b - t)Sq(t)\ .$$

If $a = b - 1$, there is no relation at hand to cancel the unfactored term. But if $a < b - 1$, we have

$$Sq(2^{b-1})Sq(2^a + 2^{b-1}) = Sq(2^a + 2^b) + \sum_{2^{b-2} \geq u \geq 1} \lambda'_u Sq(2^a + 2^b - u)Sq(u)\ .$$

Furthermore, when $u = 2^a$, then $\lambda'_u = 1$. Thus we have relations of the form:

$$\begin{aligned}&\text{For } 0 < a \le b-2,\ Sq(2^a)Sq(2^b) + Sq(2^b)Sq(2^a)\\ &+ \text{terms in the left } \mathcal{A} - \text{ideal generated by } \{Sq(1), \ldots, Sq(2^{b-1})\}\\ &= 0\,.\end{aligned}$$

The unspecified terms do not cancel the displayed terms.

We abuse notation to write $\Phi_{a,b}$ for a stable secondary operation based on a fixed choice of a defining relation.

The following table summarizes a set of defining relations for $\Phi_{i,j}$, $0 \le i \le j \le 3, j \ne i+1$ and with common domain consisting of classes annihilated by $Sq(a)$, $a = 1, 2, 4, 8$. We use the matrix notation of subsection (1.3.6). Furthermore,

$$a \ne 0 \text{ means } Sq(a)\,,$$
$$0 \text{ means } 0\,,$$
$$a.b \text{ means } Sq(a)Sq(b)\,.$$

5.3.2. Table.

degree	(i,j)						detects
1	$(0,0)$	1	0	0	0		2ι
3	$(1,1)$	3	2	0	0	1	η^2
4	$(0,2)$	4	2.1	1	0	2	2ν
7	$(2,2)$	7	6	4	0	4	ν^2
8	$(0,3)$	8	7	4.1	1	8	2σ
9	$(1,3)$	7.2	8	4.2	2		$\eta\sigma$
15	$(3,3)$	15	14	12	8		σ^2

Remark. The line for the equation of degree 8 can be replaced with

$$(8, 0, 5 + 4.1, 1)\,.$$

Remark. $\Phi_{1,3}$ detect representatives of the Toda bracket $\langle \nu, \eta, \nu \rangle$. It turns out that the stable group is $Z/2 + Z/2$, and $\eta\sigma$ is an independent element. For others, see [**67**] and [**70**].

Among the operations we have constructed, the $\Phi_{0,b}$ are the only operations of even degree.

Proposition 5.3.5. Let y be the 2-dimensional generator for the mod 2 cohomology of CP^∞. Then $\Phi_{0,b}$ is defined on y^{2^b} and

$$\Phi_{0,b}(y^{2^b}) = y^{3 \cdot 2^{b-1}} \text{ with 0 indetermanacy,}$$
$$\Phi_{0,a}(y^{2^b}) = 0 \text{ with 0 indeterminancy for } a < b\,.$$

Proof. We have

$$Sq(2i)y^k = \binom{k}{i} y^{k+i} .$$

If we write $\alpha(k)$ for the number of 1's in the dyadic expansion of k, then the action of the Steenrod algebra cannot increase the α-value of the exponent. Hence $\Phi_{0,b}$ is defined on y^{2^b} and has 0 indeterminancy. Our calculation is based on Prop. 3.5.3 for the case $K(Z,2)$. We write

$$y^{2^b} = Sq(2^b, 2^{b-1}, \ldots, 2)y .$$

Our method is to calculate $\Phi_{0,b}$ on $\Delta^* y^{2^b}$ and infer Prop. 5.3.5 by stability. We employ the notational conventions of subsection (1.3.6) in the following diagram, where we treat the case $b = 2$. In this diagram, the universal example for $\Phi_{0,2}$ on $K(9)$, and other maps, are suppressed in order to display only what matters for the calculation.

$$\begin{array}{ccccccc}
\Sigma K(Z,2) & \longrightarrow & B_2 & \longrightarrow & (\Sigma K(Z,2))^{(2)} & \xrightarrow{\Sigma p_1} & \Sigma^2 K(Z,2) \\
 & & \downarrow & & \downarrow \zeta & & \\
 & & K(Z,3) & \xrightarrow{C} & K(10,11,13) & & \\
 & & {\scriptstyle Sq(4,2)} \downarrow & & \| & & \\
 & & K(9) & \xrightarrow[\begin{bmatrix} Sq(1) \\ Sq(2) \\ Sq(4) \end{bmatrix}]{} & K(10,11,13) & \xrightarrow[{[(4),(2,1),(1)]}]{} & K(14)
\end{array}$$

where

$$C = \begin{bmatrix} Sq(1) \\ Sq(2) \\ Sq(4) \end{bmatrix} [Sq(4,2)u_3] = \begin{bmatrix} Sq(5,2) \\ 0 \\ 0 \end{bmatrix} [u_3] = \begin{bmatrix} (Sq(2)u_3)^2 \\ 0 \\ 0 \end{bmatrix}$$

by the Adem relations, the zero and squaring properties. Next, we work out ζ. From Prop. 3.5.3, we have the square

$$\begin{array}{ccc}
B_2 & \longrightarrow & (\Sigma K(Z,2))^{(2)} \\
\downarrow & & \downarrow {\scriptstyle [y|y]} \\
K(Z,3) & \xrightarrow[Sq(3)]{} & K(6) .
\end{array}$$

Since

$$(Sq(2)u_3)^2 = Sq(4)Sq(3)u_3$$

and

$$Sq(4)[y|y] = [y^2|y^2] ,$$

we obtain

$$\zeta = \begin{bmatrix} [y^2|y^2] \\ 0 \\ 0 \end{bmatrix} .$$

Now

$$[Sq(4), Sq(2,1), Sq(1)]\zeta = (\Sigma p_1)^* \begin{bmatrix} \Delta_2^* y^6 + \text{ prim.} \\ 0 \\ 0 \end{bmatrix} .$$

Since the submodule of primitives

$$PH^{12}(K(Z,2); Z/2) = 0 ,$$

we have Prop. 5.3.5 for the case $b = 2$. The general case works the same way and is left as an exercise, as is the case for $a < b$. In working through the steps, one uses the fact

$$Sq(2^a)Sq(2^b, 2^{b-1}, \ldots, 2)u_3 = 0 \text{ for } 1 \leq a \leq b .$$

Then

$$C = \begin{bmatrix} Sq(1) \\ Sq(2^{b-1}) \\ Sq(2^b) \end{bmatrix} [Sq(2^b, \ldots, 2)u_3] = \begin{bmatrix} (Sq(2^{b-1}, \ldots, 2)u_3)^2 \\ 0 \\ 0 \end{bmatrix} ,$$

$$\zeta = \begin{bmatrix} [y^{2^{b-1}}|y^{2^{b-1}}] \\ 0 \\ 0 \end{bmatrix} ,$$

$$(Sq(2^b), *, *)\zeta = (\Sigma p_1)^* \begin{bmatrix} \Delta_2^*(y^{3\cdot 2^{b-1}} + \text{ prim.}) \\ 0 \\ 0 \end{bmatrix} ,$$

and

$$PH^{3\cdot 2^b}(K(Z,2); Z/2) = 0 .$$

□

We apply Prop. 5.3.5 to answer the question raised in exercise (1.2.2).

Proposition 5.3.6. No suspension of CP^6/CP^3 is homotopy equivalent to a suspension of $S^8 \vee \Sigma^8 CP^2$.

Proof. We have a map

$$\epsilon : CP^6/CP^3 \to K(Z/2, 8)$$

such that the composite

$$CP^6 \xrightarrow{\text{proj.}} CP^6/CP^3 \xrightarrow{\epsilon} K(Z/2, 8)$$

represents y^4. Hence

$$\Phi_{0,2}(\epsilon) \neq 0 \text{ in } H^*(CP^6/CP^3)\Big/ Ind(\Phi_{0,2})\ .$$

A putative homotopy equivalence would lead to the construction of a map

$$f : CP^6/CP^3 \simeq S^8 \vee \Sigma^8 CP^2 \to S^8 \to CP^6/CP^3$$

such that

$$f \circ \epsilon \simeq \epsilon$$

and

$$\Phi_{0,2}(f \circ \epsilon) = 0$$

because the map factors through S^8. □

5.3.3. The algebraic part of the Adams decomposition formula. Let $X_0 = \mathcal{A}$ be free on one generator of degree 0. Let X_1 be the free $\mathcal{A}$-module on the generators

$$x_{1,i}, \deg x_{1,i} = 2^i,\ 0 \leq i < \infty\ .$$

Define a map

$$d : X_1 \to X_0$$

by

$$d(x_{1,i}) = \chi(Sq(2^i))$$

where χ is the canonical anti-automorphism of the Steenrod algebra. Our purpose is to determine the kernel of d.

Our method follows Shimada and Yamanoshita based on Negishi's Theorem. We proceed by first constructing certain elements in $\ker d$. These elements correspond, in the sense of (4.2.8), to the Adams operations understood in the broad sense – $\Phi_{a,b}$ is based on an Adem relation for $Sq(a)Sq(b)$. Then we can compute $\ker d$. We turn to details.

Let $\mathcal{L}(k)$ denote the left ideal in $\mathcal{A}$ spanned by the dyadic squares

$$\mathcal{L}(k) = \mathcal{A}\{Sq(1), \ldots, Sq(2^k)\}\ .$$

Let $C_1^k \subset X_1$ be the submodule generated by

$$x_{1,i},\ 0 \leq i \leq k\ .$$

Then, we have a surjective mapping

$$d : C_1^k \to \mathcal{L}(k),$$

because

$$\chi(Sq(2^i)) = Sq(2^i) + \text{ decomposable terms.}$$

Next, we observe that

$$\chi : \mathcal{L}(k) \to \mathcal{R}(k)$$

is an isomorphism, where $\mathcal{R}(k)$ is the right ideal discussed in (1.2.4).

We have the information that

$$Sq(2^k)Sq(2^i) \text{ is in } \mathcal{R}(k-1)$$

for $0 \leq i \leq k-2$ or $i = k$.

The case for $i = k$ is included in Negishi's Theorem and the other values of i are covered by inspection of the annihilator of $R(k-1)$. We call a pair of integers (i, k) *acceptable* if they obey the conditions above.

With these observations we can establish

Lemma 5.3.7. For each acceptable pair (i, k), there is an element

$$\sigma_{i,k} \text{ in } C_1^{k-1}$$

such that

$$z_{i,k} = \chi(Sq(2^i))x_{1,k} + \sigma_{i,k}$$

is a d-cycle.

Proof. We have

$$d\chi(Sq(2^i))x_{1,k} = \chi(Sq(2^k)Sq(2^i)) \; .$$

In turn, this element lies in the image of d restricted to C_1^{k-1}. □

We can now determine $\ker d$.

Proposition 5.3.8. The kernel of d restricted to C_1^k is spanned by the elements $z_{i,j}$ where (i, j) runs over acceptable pairs with $j \leq k$.

Proof. The proof is by induction on k. The case $k = 0$ is clear. Consider a d-cycle

$$z = \chi(\alpha_k)x_{1,k} + \sum_{i=0}^{k-1} \chi(\alpha_i)x_{1,i} \; .$$

Then

$$\begin{aligned} 0 &= \chi(\alpha_k)\chi(Sq(2^k)) + \sum_{i=0}^{k-1} \chi(\alpha_i)\chi(Sq(2^i)) \\ &= \chi\left(Sq(2^k)\alpha_k + \sum_{i=0}^{k-1} Sq(2^i)\alpha_i \right) \; . \end{aligned}$$

Thus

$$Sq(2^k)\alpha_k \equiv 0 \bmod \mathcal{R}(k-1) \; .$$

By Negishi's theorem, we may write

$$\alpha_k = \sum_{i=0}^{k-2} Sq(2^i)\alpha_{i,k} + Sq(2^k)\alpha'_k \; .$$

Then

$$z = \left\{ \sum_{i=0}^{k-2} \chi(\alpha_{i,k})\chi(Sq(2^i)) + \chi(\alpha'_k)\chi(Sq(2^k)) \right\} x_{1,k} + \sum_{i=0}^{k-1} \chi(\alpha_i) x_{1,i} \; .$$

Taking differences and using the elements $\sigma_{i,k}$, we have,

$$\begin{aligned} z &- \left(\sum_{i=0}^{k-2} \chi(\alpha_{i,k}) z_{i,k} \right) - \chi(\alpha'_k) z_{k,k} \\ &= \sum_{i=0}^{k-2} \chi(\alpha_{i,k})\sigma_{i,k} + \chi(\alpha'_k)\sigma_{k,k} + \sum_{i=0}^{k-1} \chi(\alpha_i) x_{1,i} \; . \end{aligned}$$

We observe that the right side of the equation is a d-cycle in C_1^{k-1}. Thus an induction on k is possible. □

Remark. $\{z_{i,j} \mid (i,j) \text{ acceptable}\}$ is a set of independent elements, their dimensions are uniquely determined by (i,j).

Remark. The results in Prop. 5.3.5 hold for secondary operations determined by $z_{0,k}$ since any Adem relation for

$$\chi(Sq(1))\chi(Sq(2^k))$$

must involve $Sq(2^k)Sq(1)$, which is the "active" part of the calculation performed in the proof. More precisely, use the relation in Prop. 5.3.3 to define $z_{0,k}$.

We turn now to the construction of the algebraic part of Adams' formula. Let X_2 be the free $\mathcal{A}$-module on the generators

$$x_2(i,j) \text{ of degree } 2^i + 2^j \text{ for acceptable } (i,j) \; .$$

Define

$$d_1 : X_2 \to X_1$$

by

$$d_1 x_2(i,j) = z_{i,j} \; .$$

Proposition 5.3.9. If $k \geq 3$, there is a d_1-cycle of the form

$$\begin{aligned} w &= \chi(Sq(1)) x_2(k,k) + \chi(Sq(2^k - 1)) x_2(1,k) \\ &+ \chi(Sq(2^k)) x_2(0,k) \\ &+ \sum c_{i,j} x_2(i,j) \end{aligned}$$

where the sum is over acceptable pairs with $j \leq k-1$.

Proof. We have, for $k \geq 3$, the Adem relation headlined as (5.3.1), and

$$\begin{aligned} d_1 w = &\ \chi(\text{relation } (5.3.1))x_{1,k} + \chi Sq(1)\sigma_{k,k} + \chi(Sq(2^k - 1))\sigma_{1,k} \\ &+ \chi(Sq(2^k))\sigma_{0,k} + \text{other terms in } C_1^{k-1} . \end{aligned}$$

Note that we need $k \geq 3$ in order to write w. Thus the summand involving the σ's is a d-cycle in $C_1'^{k-1}$. By Prop. 5.3.8, we may rewrite in terms of $z_{i,j}$ for acceptable pairs with $j \leq k - 1$. □

5.4. The Liulevicius-Shimada-Yamanoshita operations

The odd primary analogues of the Adams operations are developed in [**60**] and [**102**]. Liulevicius follows Adams in the use of homological algebra to obtain information equivalent to Prop. 5.3.9. As has been mentioned several times, Shimada and Yamanoshita make an elegant use of exact sequences from the Steenrod algebra. Toda pioneered the application of various sequences to the study of homotopy groups of spheres. The case used in [**102**] is worked out in [**86**] and we quoted the results in Chapter 1.

The homological algebra approach to the calculations is provided in [**1**], [**21**], [**60**] and the reader is urged to study those works. Moreover, the use of homological algebra is important for many applications of higher order operations. However, when sophisticated methods fail, one can still grind out a minimal resolution and then pass to the "interesting" parts. In Chapter 7 we shall encounter a fundamental connection between resolutions of certain $\mathcal{A}$-modules and factorizations of certain spaces through towers built from a resolution.

We begin the work of this section by discussing some operations based on explicit relations. In subsection (4.2.6) we looked at the Adem relation

$$(2\mathcal{P}(2), \beta\mathcal{P}(1) - 2\mathcal{P}(1)\beta)\begin{pmatrix} \beta \\ \mathcal{P}(1) \end{pmatrix} = 0 .$$

Here we mean the unsigned Bockstein. First we take the opportunity to discuss the two approaches using this relation, insertion or not of signed Bocksteins. We recast the construction from (4.2.6) with the preferred indexing of components of the stable operation. Recall that the stable Bockstein is given by

$$\beta = (\beta_n)$$

where β_n is the coefficient connecting homomorphism induced by the short exact sequence

$$0 \to Z/p \to Z/p^2 \to Z/p \to 0$$

which is obtained from

$$0 \to Z \to Z \to Z/p \to 0$$
$$1 \to (-1)^n p$$

by reduction mod p^2. The connecting homomorphism is applied to dimension n. Thus, we have

$$\beta_n : H^n(X; Z/p) \to H^{k+1}(X; Z/p) .$$

Our spaces are

$$\begin{aligned} C_{0,m} &= K(Z/p, m) , \\ C_{1,m} &= K(Z/p, m+1) \times K(Z/p, m+2p-2) , \\ C_{2,m} &= K(Z/p, m+4p-3) \text{ with } m \geq 1 . \end{aligned}$$

We have maps

$$\theta_m : C_{0,m} \to C_{1,m} \text{ given by } \theta_m = \begin{pmatrix} \beta_m \\ \mathcal{P}(1) \end{pmatrix}$$

and $\varphi_m : C_{1,m} \to C_{2,m}$ given by

$$(2\mathcal{P}(2), \beta_{m+4p-4}\mathcal{P}(1) - 2\mathcal{P}(1)\beta_{m+2p-2}) .$$

Differences in tetherings are measured by maps

$$C_{0,m} \to \Omega C_{2,m}$$

and all such factor through θ_m. Thus there is a unique stable secondary operation $\mathcal{R}$ with components

$$\mathcal{R}_m \text{ based on } [\varphi_m][\tau_m \circ \theta_m] .$$

The operation obtained with the unsigned Bockstein is

$$(-1)^m \mathcal{R}_m$$

in component m.

Proposition 5.4.1. Let u in $H^2(CP^\infty; Z/p)$ be the mod p reduction of the fundamental class. We have

(a): $\mathcal{R}(u^p) = [\![u^{3p-2}]\!]$,

(b): $\mathcal{R}(u^{p^m}) = 0, \ m > 1$.

Proof. For (a) we shall calculate

$$\mathcal{R}_{2p+1}(\Delta^* u^p)$$

and infer the result by looping once. We have $\Delta^* u^p$ represented by the composition

$$\mathcal{P}(1) \circ j_{1,\infty} : \Sigma K(Z, 2) \longrightarrow K(Z/p, 2p+1) .$$

This map appears on the left, in the following diagram

$$\begin{array}{ccccccc} \Sigma K(Z,2) & \longrightarrow & B_2 & \longrightarrow & (\Sigma K(Z,2))^{(2)} & \xrightarrow{\Sigma p_1} & \Sigma^2 K(Z,2) \\ & & \big\downarrow & & \big\downarrow{\scriptstyle \zeta} & & \\ & & K(2p+1) & \underset{\tau_{2p+1}\circ\theta_{2p+1}}{\longrightarrow} & K(2p+2,*) & & \end{array}$$

where

$$\zeta = \begin{bmatrix} +\sum_{t=1}^{p-1} \frac{1}{p} \binom{p}{t} [u^t|u^{p-t}] \\ 0 \end{bmatrix} .$$

The + sign appears because $\tau_{2p+1} = \tau$ and reverses the sign from Prop. 3.5.3. The symbol $*$ denotes $4p-1$, hence the other component of ζ is 0 for dimensional reasons. By writing out binomial coefficients and re-indexing sums, we obtain

$$\mathcal{P}(2)\zeta = (\Sigma p_1)^* \left(\frac{p-1}{2}\right) u^{3p-2} .$$

The indeterminacy is

$$2\mathcal{P}(2)H^{2p} + (\beta\mathcal{P}(1) - 2\mathcal{P}(1)\beta)H^{4p-3} = 0$$

for dimensional reasons and $\binom{p}{2} \equiv 0 \mod p$. The factor 2 in front of $\mathcal{P}(2)$ together with Prop. 3.5.2 and Theorem 4.2.4(f) give (a).

The result in (b) is proved similarly. The non-zero component in ζ is given by replacing u with

$$u^{p^{m-1}} .$$

The calculation is

$$\mathcal{R}_{2p^m+1}(\Delta^* u^{p^m}) .$$

We leave the details for an exercise. □

Next we develop an operation from an Adem relation involving

$$\beta\mathcal{P}(p^k) \text{ and } \mathcal{P}(p^k)\beta, k \geq 1 .$$

We have the Adem relation

$$[\mathcal{P}(p^k),\ \mathcal{P}(p-1)\beta, -\beta] \begin{bmatrix} \beta \\ \mathcal{P}(p^k - p + 1) \\ \mathcal{P}(p^k) \end{bmatrix} = 0 .$$

We write Ψ'_k for the stable secondary operation based on this relation, and omit the component index. If the term

$$\mathcal{P}(p^k - p + 1)$$

is rewritten in terms of

$$\mathcal{P}(1), \ldots, \mathcal{P}(p^{k-1})$$

we denote the resulting operation by Ψ_k. For these constructions we have tacitly used the signed Bockstein. Consideration of dimensions reveals that the operation obtained using the unsigned Bockstein is $(-1)^m$ times the operation above in the m-th component,

$$(-1)^m \Psi_{k,m} \; .$$

Proposition 5.4.2. Let u mean the same as in Prop. 5.4.1. We have

$$\textbf{(a):}\;\; \Psi'_k(u^{p^{k+1}}) = -[\![u^{p^k(2p-1)}]\!] \;,$$
$$\textbf{(b):}\;\; \Psi'_k(u^{p^m}) = 0 \quad m > k+1 \;,$$

with 0 indeterminacy in each case.

Proof. For (a) we calculate by substituting u^{p^k} in the entries for ζ that appear in the diagram for Prop. 5.4.1. The "active" part of the calculation involves the term from the Adem relation,

$$\mathcal{P}(p^k)\beta \; .$$

We leave the details as an exercise.

For (b) we calculate

$$\Psi'_k(\Delta^* u^{p^m})$$

and substitute $u^{p^{m-1}}$ for the entries of ζ. □

The term "basic" is often used to describe the Adams operations. It has precise meaning in terms Prop. 5.3.8 — any d-cycle is an $\mathcal{A}$-linear combination of the elements $z_{i,j}$. Consequently, any stable secondary operation defined on the intersection

$$\bigcap_{i \geq 0} \ker Sq(2^i)$$

is an $\mathcal{A}$-linear combination of Adams operations. Next we summarize the corresponding results for odd primes. In the following table, the notation for d-cycles is taken from [**102**]. Next to each entry is Liulevicius name for the operation. The reader who looks into the details of these papers will find that Liulevicius is naming operations in terms of the cohomology of the Steenrod algebra.

cycle	relation	Liulevicius name
$k \geq 1$		
$Z_{i,k}$	$\chi(\mathcal{P}(p^i))\chi(\mathcal{P}(p^k)) + \sigma_{i,k}\ 0 \leq i \leq k-2$	$h_i h_k$
$Z_{-1,k}$	$\chi(\beta)\chi(\mathcal{P}(p^k)) - \chi(\beta\mathcal{P}(p^k-1))\chi(\mathcal{P}(1)) - \chi(\mathcal{P}(p^k))\chi(\beta)$	$a_0 h_k$
U_k	$\chi(\mathcal{P}(2p^{k-1}))\chi(\mathcal{P}(p^k)) + \sigma'_k$	μ_k
V_k	$\chi[2\mathcal{P}(p^k + p^{k-1}) - \mathcal{P}(p^k)\mathcal{P}(p^{k-1})]\chi(\mathcal{P}(p^k)) + \sigma''_k$	ν_k
W_k	$\chi(\mathcal{P}(p^k(p-1))\chi(\mathcal{P}(p^k)) + \sigma'''_k$	λ_k
and for		
$k=0$		
V_0	$\chi(2\beta\mathcal{P}(1) - \mathcal{P}(1)\beta)\chi(\mathcal{P}^1) - 2\chi(\mathcal{P}(2))\chi(\beta)$	ρ
W_0	$\chi(\mathcal{P}(p-1))\chi(\mathcal{P}(1))$	λ_0
and for		
$k=-1$		
$Z_{-1,-1}$	$\chi(\beta)\chi(\beta)$	$a_0 a_0$

Our $\mathcal{R}$ corresponds to V_0 and Ψ_k to $Z_{-1,k}$, up to possible non-zero multiples from Z/p.

For other uses, we write out the low indexed cases.

name	relation
$a_0 a_0$	β^2
λ_0	$\mathcal{P}(p-1)\mathcal{P}(1) = 0$
μ_0	$[\mathcal{P}(p,1) - 2\mathcal{P}(p+1), 2\mathcal{P}(2)] \begin{bmatrix} \mathcal{P}(1) \\ \mathcal{P}(p) \end{bmatrix} = 0$
ν_0	$[2\mathcal{P}(2p), \mathcal{P}(p+1) - 2\mathcal{P}(p,1)] \begin{bmatrix} \mathcal{P}(1) \\ \mathcal{P}(p) \end{bmatrix} = 0$

where $\mathcal{P}(a,b)$ means $\mathcal{P}(a)\mathcal{P}(b)$.

The notation in the table distinguishes among terms in the kernel of left multiplication by $\mathcal{P}(p^k)$. Cycles denoted as $Z_{*,*}$ arise in the same way as the cycles discussed in the previous section. The other cycles arise from those terms in Mukohda's theorem which involve certain decomposable elements of the Steenrod algebra.

The algebraic form of the Liulevicius-Shimada-Yamanoshita decomposition formula is developed in [**102**] by the method discussed in our previous section. The results are d_1-cycles of the form

$$w = \beta[W_0] + \mathcal{P}(p-2)[V_0]$$

and for $k \geq 1$,

$$\begin{aligned} w = {} & \chi(\beta)[W_k] - \chi(\beta\mathcal{P}(p^k(p-1)-1)[Z_{0,k}] \\ & - \chi(\mathcal{P}(p^k(p-1)))[Z_{-1,k}] \\ & + \text{ terms involving } \{Z_{i,j}, U_j, V_j, W_j; j < k\} , \\ & \text{with } k = 1, \text{ substitute } \chi(\beta\mathcal{P}(p(p-1)-2))[U_1] \text{ for the term on } [Z_{0,1}] . \end{aligned}$$

Here square brackets take over the role of $x_2(i,j)$ in the previous section.

Finally, we remark that the calculation in Prop. 5.4.2 depends on the presence of

$$\mathcal{P}(p^k)\beta$$

in the Adem relation. Hence operations based on $Z_{-1,k}$ give the same results as stated for Ψ'_k, up to non-zero multiple from Z/p.

5.5. Operations based on unstable relations

We study operations based on Adem relations of the form

$$P = \sum_{t=1}^{T} b_t a_t \text{ with } \deg b_t,\ \deg a_t > 0 ,$$

where P is an unfactored primary operation. There are three principal examples:

(1): $P = Sq(n+1), n+1 \neq 2^i$,

(2): $P = \beta\mathcal{P}(n)$,

(3): $P = \mathcal{P}(n),\ n \neq p^i$.

We write

$$\begin{aligned} \theta = {} & \text{ col } (a_1, \ldots, a_T) , \\ \varphi = {} & (b_1, \ldots, b_T) , \end{aligned}$$

and express

$$P = \Omega\varphi \circ \Omega\theta .$$

Write

$$\Omega C_0 = K(Z/p, m)$$

where, in the various cases,

(1): $p = 2,\ m = n$,

(2): p odd prime, $m = 2n$,

(3): p odd prime, $m = 2n - 1$.

Write

$$\Omega C_1 = \prod_{t=1}^{T} K(Z/p, m + |a_t|) \qquad |a_t| = \text{ degree } a_t \ ,$$

$$\Omega C_2 = K(Z/p, m + |P|)$$

with maps

$$\Omega C_0 \xrightarrow{\Omega\theta} \Omega C_1 \xrightarrow{\Omega\varphi} \Omega C_2 \ .$$

Thus, in each case

$$\Omega\varphi \circ \Omega\theta \sim *$$

but

$$\varphi \circ \theta \not\sim *$$

and we are set to use the results in subsections (3.5.7) and (4.2.9).

Before getting into detail, we look at the literature. The operation described by (1) was first studied by Kristensen [**55**] and Brown and Peterson [**17**]. It was then used in a study of the celebrated Kervaire invariant [**51**]. This study appears as [**18**]. We shall develop two aspects of the Kristensen-Brown-Peterson operation, its use in detecting Whitehead products and homology squares.

The operation described in (2) was first studied by Zabrodsky [**118**]. Its principal feature is that it detects homology p-th powers. It is a central tool in the study of finite H-spaces.

The operation described in (3) is well known in an incidental way. It detects the first unstable element at odd primes in the homotopy groups of odd dimensional spheres. Here, we use it to study a point in the relation between cogroups and suspensions, to refine a result in [**12**].

We introduce the notation for our operations. They are generically associated to

$$(\Omega\varphi, \tau \circ \Omega\theta, H)$$

for some tethering H, but we use the loop space as in subsection (4.2.9) for the universal example.

(1): $S_{\Omega\theta}(X) = \{x \text{ in } H^n(X; Z/2) \mid a_t x = 0\}$
$T_{\Omega^2\varphi}(X)$ is a quotient of $H^{2n}(X; Z/2)$,

$$K_n : S_{\Omega\theta}(\quad) \to T_{\Omega^2\varphi}(\quad) \ .$$

(2): $S_{\Omega\theta}(X) = \{x \text{ in } H^{2n}(X : Z/p) \mid a_t x = 0\}$
$T_{\Omega^2\varphi}(X)$ is a quotient of $H^{2np}(X : Z/p)$,

$$Z_n : S_{\Omega\theta}(\quad) \to T_{\Omega^2\varphi}(\quad) \ .$$

(3): $S_{\Omega\theta}(X) = \{x \text{ in } H^{2n-1}(X : Z/p) \mid a_t x = 0\}$
$T_{\Omega^2\varphi}(X)$ is a quotient of $H^{2np-2}(X : Z/p)$,

$$\Gamma_n : S_{\Omega\theta}(\quad \to T_{\Omega^2\varphi}(\quad) .$$

We now investigate the Kristensen-Brown-Peterson operation.

Proposition 5.5.1. Let u in $H^n(\Omega S^{n+1}; Z/2)$ be the generator. Then K_n is defined on u and $K_n(u)$ is the non-zero class in $H^{2n}(\Omega S^{n+1}; Z/2)$ with 0 indeterminancy.

Proof. We set up the diagram

$$\begin{array}{ccccc}
S^{n+1} & = & S^{n+1} & & \\
\downarrow{\scriptstyle \bar{\epsilon}} & & \downarrow{\scriptstyle \epsilon} & & \\
W_\theta & \xrightarrow{p} & K(n+1) & \xrightarrow{\theta} & C_1 \\
\downarrow{\scriptstyle h} & & \| & & \downarrow{\scriptstyle \varphi} \\
W & \xrightarrow[q]{} & K(n+1) & \xrightarrow[Sq(n+1)]{} & K(2n+2)
\end{array}$$

with $\bar{\epsilon}$ satisfying

$$p\bar{\epsilon} \simeq \epsilon .$$

We choose ϵ so that $\Omega\epsilon$ represents u. Then looping yields a composition

$$\Omega S^{n+1} \xrightarrow{\Omega\bar{\epsilon}} \Omega W_\theta \xrightarrow{\Omega h} \Omega W \xrightarrow{\gamma^{\natural\natural}} K(2n)$$

where the notation is from Prop. 3.5.5. Then the Kristensen-Brown-Peterson operation is given by

$$K_n(u) = [\![\gamma^{\natural\natural} \circ \Omega h \circ \Omega\bar{\epsilon}]\!] .$$

This value in $T_{\Omega^2\varphi}(\Omega S^{n+1})$ does not depend on possible choices of the lifting $\bar{\epsilon}$.

We compute its value by diagram chasing between the cohomology Serre spectral sequence for the path-loop fibrations over S^{n+1} and over W. Thus at E_2 we have

$$(h\bar{\epsilon})^* \otimes (\Omega h\Omega\bar{\epsilon})^* : H^*W \otimes H^*\Omega W \to H^*S^{n+1} \otimes H^*\Omega S^{n+1}$$

and we have observed a homotopy equivalence,

$$\Omega W \to \Omega W \times \Omega W \xrightarrow{\Omega q \times \gamma^{\natural\natural}} K(n) \times K(2n) .$$

We now observe that in H^*W,

$$q^* b_{n+1}^2 = q^* Sq(n+1) b_{n+1} = 0,$$

where b_{n+1} is the fundamental class of $K(n+1)$. Then diagram chasing yields Prop. 5.5.1. In particular, $E_2 = E_n$ for dimensional reasons, and

$$d_n u = u' \, ,$$
$$d_n \Omega q^* b_n = q^* b_{n+1}$$

with u' represented by ϵ. Hence, by naturality, the map between spectral sequences sends

$$m_1 \doteq q^* b_{n+1} \otimes \Omega q^* b_n$$

to

$$m_2 \doteq u' \otimes u$$

which in turn is the image under d_n of the $2n$-dimensional generator of the cohomology of ΩS^{n+1}. Since

$$d_n(m_1) = q^* b_{n+1}^2 = 0$$

it follows that m_1 is a d_n-boundary of an element mapping non-trivially to the cohomology of ΩS^{n+1}. Now in terms of the splitting of ΩW, all the $2n$-dimensional elements from $K(n)$ are d_n-cycles. This leaves only

$$(\gamma^{\natural\natural})^* b_{2n}$$

to be mapped non-trivially by d_n. Hence it is also mapped non-trivially by

$$(\Omega h \Omega \bar{\epsilon})^* \, .$$

□

It follows that K_n detects the Whitehead product

$$[\iota_n, \iota_n] \text{ in } \pi_{2n-1} S^n, \ n \neq 2^i - 1$$

since this element serves as the first attaching map in a cell decomposition of ΩS^{n+1}.

The operations K_n for n even and $n \equiv 1 \mod 4$ are related to those studied in the first section of this chapter, exercise (5.1.8):

$$K_{2n} = \Psi_{2n} + \text{ primary operation} \qquad (5.1.8a),$$
$$\Omega K_{4n+1} = \Phi_{4n} + \text{ primary operation} \qquad (5.1.8b),$$

if the same Adem relation is used on both sides of each equation.

We combine these two facts with the calculations of (5.1.8) to obtain

$$[\iota_{2n}, \iota_{2n}] \text{ is not in the image of suspension}$$

and

$$[\iota_{4n+1}, \iota_{4n+1}] \text{ is not in the image of double suspension.}$$

If the Whitehead product for ι_{2n} were the suspension of a class α, then we have a map

$$\Sigma(S^{2n-1} \cup_\alpha e^{4n-1}) \xrightarrow{f} \Omega S^{2n+1}$$

such that

$$f^*u = \Delta^* v, \quad |v| = 2n - 1$$

and $f^* \neq 0$ in dimension $4n$. But

$$f^* K_{2n}(u) = \Psi_{2n}(\Delta^* v) + f^* \delta u$$

where δ is a primary operation. Since both

$$\delta u = 0 \,,$$
$$[\![\Delta^*(v \cup Sq(1)v)]\!] = 0 \,,$$

f is an impossible map. The argument for the other Whitehead product is similar.

Remark. The method can be applied to other spaces. For example, the space Y obtained as a push-out,

$$\begin{array}{ccc} S^5 & \longrightarrow & S^5 \cup_{[\iota,\iota]} e^{10} \\ \big\downarrow & & \big\downarrow \\ S^5 \cup_{\eta^2} e^8 & \longrightarrow & Y \end{array}$$

is not homotopy equivalent to a double suspension.

Remark. If η^2 were replaced by η, we receive $\Sigma^2 SU(3)$ as a push-out, up to homotopy.

We now turn to squares in homology. We may regard the space described in the proof of Prop. 5.5.1 as the universal example for K_n,

$$\begin{array}{ccccc} \Omega W_\theta & \xrightarrow{\widetilde{\Omega\varphi}} & K(2n) & & \\ \big\downarrow & & & & \\ K(n) & \xrightarrow[\Omega\theta]{} & \Omega C_1 & \xrightarrow[\Omega\varphi]{} & K(2n+1) \end{array}$$

where we have abused the notation of (3.5.7) and written

$$\widetilde{\Omega\varphi} \text{ for } \gamma^{\natural\natural} \circ \Omega h \,.$$

The elements γ and ζ featured in Prop. 3.5.4 are worked out in the diagram,

$$\begin{array}{ccccc}
B_2W & \longrightarrow & (\Sigma\Omega W)^{(2)} & \xrightarrow{\Sigma p_1} & \Sigma^2\Omega W \\
\downarrow & & \downarrow{\scriptstyle (\Sigma\Omega q)^{(2)}} & & \\
B_2 & \longrightarrow & (\Sigma K(n))^{(2)} & & \downarrow{\scriptstyle \gamma} \\
\downarrow & & \downarrow{\scriptstyle \zeta} & & \\
K(n+1) & \xrightarrow[Sq(n+1)]{} & K(2n+2) & = & K(2n+2)
\end{array}$$

$$\zeta = [b_n|b_n] \; ,$$
$$\gamma = \Delta_2^* b_{2n} + \text{ terms from } \Delta_2^* H^{2n} K(n) \; .$$

Since the ambiguous terms are primitive, they do not affect the value of the composition

$$\gamma \circ \Sigma p_1 \; .$$

Writing the result in the notation of cohomology, we have

$$(\gamma \circ \Sigma p_1)^*(b_{2n+2}) = [\Omega p^* b_n | \Omega p^* b_n] \; .$$

Let

$$c_n \text{ in } H_n(\Omega W_\theta; Z/2)$$

satisfy

$$\langle c_n, \Omega p^* b_n \rangle \neq 0$$

under the Kronecker pairing. Then, we have the calculation

$$\begin{aligned}
\langle c_n^2, \widetilde{\Omega\varphi}^* b_{2n} \rangle &= \langle c_n \otimes c_n, \; \Omega p^* b_n \otimes \Omega p^* b_n \rangle \\
&= \langle c_n, \Omega p^* b_n \rangle^2 \\
&\neq 0 \; .
\end{aligned}$$

If Y is an H-space, and $f : Y \to K(2n)$ is an H-map, which lifts to an H-map

$$\bar{f} : Y \to \Omega W_\theta \; ,$$

then we can pass this calculation to the cohomology of Y by naturality. The useful applications usually do not come with so much structure for f, and there is an extensive theory to deal with this situation [**57**], [**119**], [**48**].

We next discuss the non-additivity property of K_n.

Proposition 5.5.2. Let u, v in $H^n(X; Z/2)$ be in the domain of K_n for a space X. Then K_n is defined on the sum $u + v$ and

$$K_n(u+v) = K_n(u) + K_n(v) + [\![u \cup v]\!] \; .$$

Proof. Let

$$\epsilon_1, \epsilon_2 : X \to K(n)$$

represent u, v respectively, and

$$\bar{\epsilon}_1, \bar{\epsilon}_2 : X \to \Omega W_\theta$$

be liftings into the universal example for K_n. Then we have the composition

$$X \xrightarrow{\Delta} X \times X \xrightarrow{\bar{\epsilon}_1 \times \bar{\epsilon}_2} \Omega W_\theta \times \Omega W_\theta \xrightarrow{m} \Omega W_\theta$$

which is a lifting of a map representing the sum $u + v$. Then

$$\begin{aligned} K_n(u+v) &= [\![\widetilde{\Omega\varphi} \circ m \circ \bar{\epsilon}_1 \times \bar{\epsilon}_2 \circ \Delta]\!] \\ &= [\![(\bar{\epsilon}_1 \times \bar{\epsilon}_2 \circ \Delta)^*((\Omega p)^* b_n \otimes (\Omega p)^* b_n + 1 \otimes \widetilde{\Omega\varphi}^* b_{2n} \\ &\qquad + \widetilde{\Omega\varphi}^* b_{2n} \otimes 1)]\!] \\ &= K_n(u) + K_n(v) + [\![\Delta^*(\epsilon_1^* b_n \otimes \epsilon_2^* b_n)]\!] \\ &= K_n(u) + K_n(v) + [\![u \cup v]\!] \,. \end{aligned}$$

□

Exercise 5.5.3. Derive Prop. 5.5.1 from Prop. 5.5.2. This is the method in [**17**].

Remark. The properties of K_n described in Prop. 5.5.1 and Prop. 5.5.2 do not depend on the choice of factorization for $Sq(n+1)$. We leave as an exercise with the Adem relations, that if $(n+1)$ is not a power of 2, then $Sq(n+1)$ has a factorization where each b_t in (1) has degree greater than 1, provided $n \geq 4$. Now, by means of diagram chasing in Serre spectral sequences for path-loop fibrations involving the 2-local EHP sequence

$$S^n \xrightarrow{E} \Omega S^{n+1} \xrightarrow{H} \Omega S^{2n+1} \,,$$

it is possible to show that the evaluation map

$$\Sigma\Omega^2 S^{n+1} \to \Omega S^{n+1}$$

is surjective in mod 2 homology in dimensions of the form n times a power of 2. Thus the mod 2 homology of ΩS^{n+1} has a simple system of transgressive generators in the homology spectral sequence for the path-loop fibration over ΩS^{n+1}. By Borel's Theorem, we obtain that the mod 2 homology of $\Omega^2 S^{n+1}$ is a polynomial algebra

$$H_*(\Omega^2 S^{n+1}; Z/2) = Z/2[x_0, x_1, x_2 \ldots]$$

where degree $|x_k| = 2^k n - 1$, $k \geq 0$. Moreover, from Prop. 5.5.1 it follows that ΩK_n operates isomorphically from

$$H^{n-1}(\Omega^2 S^{n+1}) \to H^{2n-1}(\Omega^2 S^{n+1})$$

provided $n \neq 2$ or $n+1$ is not a power of 2 and $Sq(n+1)$ is factored as described above. Calculation with integral coefficients reveals that

$$Sq(1) : H^{2n-2}(\Omega^2 S^{n+1}) \to H^{2n-1}(\Omega^2 S^{n+1})$$

is an isomorphism, so the choice of factorization matters. The reader who wishes to explore the connection of our material with the Araki-Kudo-Dyer-Lashof operations in homology should look at [**26**]. I also suggest a paper by Goerss and Mahowald [**35**] where the K-B-P operation is used.

The next topic is the Zabrodsky operation Z_n. Unlike the situation for the K-B-P operation, there is a preferred choice of null-homotopy for the composition

$$\Sigma\Omega K(2n+1) \xrightarrow{j_{1,\infty}} K(2n+1) \underset{\beta\mathcal{P}(n)}{\longrightarrow} K(2np+2)\ .$$

We make the choice such that ζ is given by the formula in Prop. 3.5.3(b). We have the diagram

$$\begin{array}{ccccc}
W_\theta & \xrightarrow{\ p\ } & K(2n+1) & \xrightarrow{\ \theta\ } & C_1 \\
{\scriptstyle h}\downarrow & & \| & & \downarrow{\scriptstyle \varphi} \\
W & \xrightarrow[q]{} & K(2n+1) & \xrightarrow[\beta\mathcal{P}(n)]{} & K(2np+1)\ .
\end{array}$$

Then the universal example for Z_n and key information is displayed in the diagram

$$\begin{array}{ccccc}
 & & (\Sigma\Omega W_\theta)^{(2)} & \xrightarrow{\ \Sigma p_1\ } & \Sigma^2\Omega W_\theta \\
 & & \downarrow{\scriptstyle (\Sigma\Omega h)^{(2)}} & & \\
B_2 W & \longrightarrow & (\Sigma\Omega W)^{(2)} & & \downarrow{\scriptstyle \gamma\circ\Sigma^2\Omega h} \\
\downarrow & & \downarrow & & \\
B_2 K(2n+1) & \longrightarrow & (\Sigma K(2n))^{(2)} & & \\
\downarrow & & \downarrow{\scriptstyle \zeta} & & \\
K(2n+1) & \xrightarrow[\beta\mathcal{P}(n)]{} & K(2np+2) & = & K(2np+2)
\end{array}$$

with

$$\zeta = -\sum_{t=1}^{p-1} \frac{1}{p}\binom{p}{t}\left[b_{2n}^t \mid b_{2n}^{p-t}\right]\ .$$

As in the previous example, we calculate a non-zero p-th power in the homology of the universal example. Let

$$c_{2n} \text{ in } H_{2n}(\Omega W_\theta; Z/p)$$

satisfy

$$\langle c_{2n}, (\Omega p)^* b_{2n} \rangle = 1$$

under the Kronecker pairing. Then $c_{2n}^p \neq 0$. In fact,

$$\langle c_{2n}^p, \widetilde{\Omega\varphi}^* b_{2np} \rangle = 1$$

where $\widetilde{\Omega\varphi}$ represents Z_n.

We leave the details as an exercise after remarking that

$$\langle c_{2n}^{p-1}, \ (\Omega p)^* b_{2n}^{p-1} \rangle \equiv (p-1)!$$

which is congruent to -1 mod p by Wilson's Theorem. This minus sign cancels that in ζ.

The reader will find many applications of Zabrodsky's operation in the previously mentioned references to finite H-space theory. We turn to a different application of Zabrodsky's calculation to prove a theorem of J. Moore [**83**]. In Prop. 5.5.4 below n is any positive integer.

Proposition 5.5.4. As Hopf algebras over Z/p (p an odd prime)

$$H_*(\Omega^2 S^{2n+1}; Z/p) \cong \Lambda(x_{2n-1}) \otimes \bigotimes_{t=1}^{\infty} Z/p[x_{d(t)}] \otimes \Lambda(y_{e(t)})$$

where $d(t) = 2np^t - 2$, $e(t) = 2np^t - 1$ and denote dimension. Furthermore, the generators are primitive and the homology Bockstein connecting homomorphism satisfies

$$\beta y = x$$

for each generator.

Before giving the proof, we recall the two fibration sequences of classical homotopy theory (for p-local spaces),

$$\text{James:} \qquad J_{p-1}S^{2n} \xrightarrow{j} \Omega S^{2n+1} \xrightarrow{H} \Omega S^{2np+1},$$

$$\text{Toda:} \qquad S^{2n-1} \xrightarrow{\Omega j} \Omega J_{p-1}S^{2n} \xrightarrow{T} \Omega S^{2np-1}.$$

The treatment in [**84**] is recommended for background.

The calculation for Prop. 5.5.4 is based on the following sequence of fibrations, obtained by fitting James fibrations together.

$$\begin{array}{ccccccc}
\Omega J_{p-1}S^{2n} & & \Omega J_{p-1}S^{2pn} & & \Omega J_{p-1}S^{2p^2n} & & \\
\downarrow & & \downarrow & & \downarrow & & \\
\Omega^2 S^{2n+1} & \longrightarrow & \Omega^2 S^{2pn+1} & \longrightarrow & \Omega^2 S^{2p^2n+1} & \longrightarrow & \cdots
\end{array}$$

The strategy is to show that each fibration satisfies the condition, *totally non-homologous to zero mod p*.This means that the inclusion of the fiber is an injection in mod p homology, hence the Serre spectral sequence collapses. We also import the information

$$H_*(\Omega J_{p-1}S^{2m}; Z/p) = \Lambda(x_{2m-1}) \otimes Z/p[x_{2mp-2}]$$

as Hopf algebras with primitive generators.

There are several steps in our proof of Prop. 5.5.4.

Lemma 5.5.5. p-th powers in the mod p cohomology of $\Omega^2 S^{2m+1}$ are 0.

Proof. We use a result of Browder [**14**] on the homology suspension for H-spaces X,

$$\sigma_* : Q_k(\Omega X; Z/p) \to PH_{k+1}(X : Z/p)$$

is monic provided $k \not\equiv -2 \bmod 2p$. Since the first non-zero p-th power in cohomology is primitive and the homology of ΩS^{2m+1} is concentrated in even dimensions, the lemma follows. □

In particular, a general result from [**82**] can be invoked to conclude that the homology of $\Omega^2 S^{2m+1}$ is primitively generated.

Lemma 5.5.6. The map on mod p homology induced by inclusion of the fiber

$$H_{2mp-2}(\Omega J_{p-1}S^{2m}) \to H_{2mp-2}(\Omega^2 S^{2m+1})$$

is a monomorphism.

This fact follows from examination of the Serre spectral sequence in homology for the path-loop fibrations

$$\begin{array}{ccc} \Omega J_{p-1}S^{2m} & \longrightarrow & \Omega^2 S^{2m+1} \\ \downarrow & & \downarrow \\ * & \longrightarrow & * \\ \downarrow & & \downarrow \\ J_{p-1}S^{2m} & \longrightarrow & \Omega S^{2m+1} \end{array}$$

using Z/p and $Z_{(p)}$ coefficients. The calculation also shows

$$H_{2mp-2}(\Omega S^{2m+1}; Z_{(p)}) = Z/p$$

and hence the first Bockstein in Prop. 5.5.4. □

We now use the Zabrodsky operations in Prop. 5.5.4. We have

$$Z_{np-1} \quad \text{based} \quad \beta\mathcal{P}(np-1)\ .$$

We bring in Toda's map T in the diagram,

$$(1)\qquad \begin{array}{ccccc}
\Omega S^{2np-1} & & \xrightarrow{\quad\bar{\epsilon}_1\quad} & \Omega W_\theta & \xrightarrow{\tilde{\beta}} K(2p(np-1)) \\
\uparrow{\scriptstyle T} & & & \downarrow & \\
\Omega J_{p-1}S^{2n} & \xrightarrow[\Omega j]{} & \Omega^2 S^{2n+1} & \xrightarrow[\epsilon_2]{} & K(2np-2)
\end{array}$$

where $\bar{\epsilon}_1$ is a lift of

$$\epsilon_1 : \Omega S^{2np-1} \to K(2np-2)$$

and the maps ϵ_1 and ϵ_2 represent the bottom dimensional classes so that

$$\epsilon_2 \circ \Omega_j \simeq \epsilon_1 \circ T\ .$$

Let $\bar{\epsilon}_2$ be a lift of ϵ_2. Since the cohomology of $\Omega J_{p-1}S^{2n}$ is 0 in dimension

$$2p(np-1)-1$$

we obtain

$$\bar{\epsilon}_1 \circ T \simeq \bar{\epsilon}_2 \circ \Omega j\ .$$

Since $\bar{\epsilon}_1$ is an H-map, and the Zabrodsky operation detects p-th powers, we have

$$Z_{np-1}(\epsilon_1 \circ T) \neq 0\ .$$

Hence Ωj is an injection in homology on $x^k_{d(1)}$, $1 \leq k \leq p$. Furthermore, the first non-zero element in

$$\ker(\Omega j)*$$

is primitive, so products of these classes with x_{2n-1} are mapped monomorphically.

We now work up inductively through the p-th powers of $x_{d(1)}$. We have the diagram for the Zabrodsky operation

$$Z_{p(np-1)}$$

accepting maps as in the following diagram, and we use the James-Hopf map as part of the construction,

$$(2)\qquad \begin{array}{ccccc}
\Omega S^{2np-1} \xrightarrow{H} & \Omega S^{2p(np-1)+1} & \xrightarrow{\bar{\epsilon}'_1} & \Omega W_\theta & \longrightarrow K(2p^2(np-1)) \\
\uparrow{\scriptstyle T} & & & \downarrow & \\
\Omega J_{p-1}S^{2n} & \xrightarrow[\Omega j]{} \Omega S^{2n+1} & \xrightarrow[\epsilon'_2]{} & K(2p(np-1)) &
\end{array}$$

where $\epsilon_2' = \tilde{\boldsymbol{\beta}} \circ \bar{\epsilon}_2$ from (1), the map $\vec{\epsilon}_1'$ is an H-map lifting

$$\epsilon_1' : \Omega S^{2p(np-1)+1} \to K(2p(np-1))$$

such that $\epsilon_1' \circ H \circ T$ represents $x^p_{d(1)}$. Thus

$$Z_{p(np-1)}(\epsilon_1' \circ H \circ T) \neq 0 \, ,$$

and for any lift $\vec{\epsilon}_2'$ of ϵ_2', we have

$$\vec{\epsilon}_2' \circ \Omega j \simeq \vec{\epsilon}_1' \circ H \circ T$$

since the cohomology of their source is 0 in dimension

$$2p^2(np-1) - 1 \, .$$

The proof that loops on the James fibration is totally non-homologous to zero mod p now follows by routine induction. The proof of Prop. 5.5.4 is completed by using this fact inductively in the sequence of fibrations

$$\begin{array}{ccccc}
\Omega J_{p-1} S^{2n} & & \Omega J_{p-1} S^{2np} & & \\
\downarrow & & \downarrow & & \\
\Omega^2 S^{2n+1} & \longrightarrow & \Omega^2 S^{2np+1} & \longrightarrow & \Omega^2 S^{2np^2+1} \quad \cdots \, .
\end{array}$$

□

The final topic in this section is the unstable operation Γ_n based on a factorization of $\mathcal{P}(n)$. As with the other unstable operations, we produce its universal example in a path-loop fibration. Thus if

$$\mathcal{P}(n) = \Omega\varphi \circ \Omega\theta \, ,$$

we write the diagrams,

$$\begin{array}{ccccc}
W_\theta & \xrightarrow{p} & K(2n) & \xrightarrow{\theta} & C_1 \\
{\scriptstyle h}\downarrow & & \| & & \downarrow{\scriptstyle \varphi} \\
W & \xrightarrow[q]{} & K(2n) & \longrightarrow & K(2np)
\end{array}$$

and

$$\begin{array}{ccc}
\Omega W_\theta & \xrightarrow{\widetilde{\Omega\varphi}} & K(2np-2) \\
{\scriptstyle \Omega p}\downarrow & & \\
K(2n-1) & \xrightarrow[\Omega\theta]{} & \Omega C_1
\end{array}$$

where $\widetilde{\Omega\varphi}$ is really the double adjoint of a map made using Prop. 3.5.3(c) in the diagram below,

$$\begin{array}{ccccc}
 & & & & \Sigma^2\Omega W_\theta \\
 & & & & \Big\downarrow {\scriptstyle j_{1,p-1}\circ\Sigma^2\Omega h} \\
B_pW & \longrightarrow & (\Sigma\Omega W)^{(p)} & \longrightarrow & \Sigma B_{p-1}W \\
\Big\downarrow & & \Big\downarrow & & \Big| \\
B_p & \longrightarrow & (\Sigma K(2n-1))^{(p)} & & \Big| {\scriptstyle \gamma} \\
\Big\downarrow & & \Big\downarrow {\scriptstyle \zeta} & & \Big\downarrow \\
K(2n) & \underset{\mathcal{P}(n)}{\longrightarrow} & K(2np) & = & K(2np)
\end{array} .$$

Thus $\Omega W \simeq K(2n-1)\times K(2np-2)$ as H-spaces with the product structure. In particular, Γ_n is additive.

We can make a calculation that is virtually identical with Prop. 5.5.1. Throughout this discussion, $n \neq p^i$ unless otherwise stipulated.

Proposition 5.5.7. Let u in $H^{2n-1}(\Omega J_{p-1}S^{2n}; Z/p)$ be a non-zero element. Then Γ_n is defined on u and $\Gamma_n(u)$ is a non-zero class in $H^{2np-2}(\Omega J_{p-1}S^{2n}; Z/p)$ with 0 indeterminacy.

Proof. We may construct the diagram

$$\begin{array}{ccccc}
J_{p-1}S^{2n} & = & J_{p-1}S^{2n} & & \\
\Big\downarrow {\scriptstyle \bar{\epsilon}} & & \Big\downarrow {\scriptstyle \epsilon} & & \\
W_\theta & \overset{p}{\longrightarrow} & K(2n) & \overset{\theta}{\longrightarrow} & C_1 \\
\Big\downarrow {\scriptstyle h} & & \Big\| & & \Big\downarrow {\scriptstyle \varphi} \\
W & \underset{q}{\longrightarrow} & K(2n) & \underset{\mathcal{P}(n)}{\longrightarrow} & K(2np)
\end{array}$$

with $\Omega\epsilon$ representing u. Then Γ_n is defined on u because Steenrod operations on $J_{p-1}S^{2n}$ are 0, and

$$\Gamma_n(u) = [\![\widetilde{\Omega\varphi} \circ \Omega h \circ \Omega\bar{\epsilon}]\!] .$$

The indeterminacy is 0 for dimensional reasons. The argument that $\Gamma_n(u)$ is non-zero is a reprise of the argument in Prop. 5.5.1. The corresponding facts are, besides the splitting of ΩW,

$$(q^* b_{2n})^p = 0 \ .$$

By a similar diagram chase, we obtain the information that in the spectral sequence for the path-loop fibration over W,

$$q^* b_{2n}^{p-1} \otimes \Omega q^* b_{2n-1}$$

is a $d_{2n(p-1)-1}$-boundary of an element from ΩW which is mapped non-trivially to the cohomology of $\Omega J_{p-1} S^{2n}$ by $(\Omega h \circ \Omega \bar{\epsilon})^*$. The terms from

$$H^{2np-2}(K(2n-1))$$

are mapped to 0 because they are in the ideal generated by primary operations on b_{2n-1} and primary operations vanish on u. Thus

$$\widetilde{\Omega\varphi}^* b_{2np-2}$$

is the only element that can be so mapped. □

One consequence of Prop. 5.5.7 is that Γ_n detects the first attaching map in a cell decomposition for $\Omega J_{p-1} S^{2n}$. We can use Prop. 5.5.4 (only the information for $t \leq 1$ is needed) to gain some information about this map. First we lift the restriction that n not be a power of p as Prop. 5.5.4 holds without this restriction. We look at the homotopy sequence for the p-local pair

$$(\Omega^2 S^{2n+1}, \ S^{2n-1}),$$

$$\pi_{2np-2}(\Omega^2 S^{2n+1}, \ S^{2n-1}) \xrightarrow{\partial} \pi_{2np-3}(S^{2n-1}) \xrightarrow{E^2} \pi_{2np-3}(\Omega^2 S^{2n+1}) \ .$$

The group on the left is Z/p, by the Hurewicz Theorem. Write

$$C_{2np-2} : (D^{2np-2}, \dot{D}^{2np-2}) \to (\Omega^2 S^{2n+1}, S^{2n-1})$$

for a generator, and define

$$w_n \doteq \partial C_{2np-2} \ .$$

Thus w_n has order $\leq p$ and generates $\ker E^2$. When $n \neq p^i$, then Prop. 5.5.7 yields

$$w_n \neq 0 \ .$$

Next, we look at a particular Adem relation and prove that w_n, for certain values of n, is not in the image of the double suspension homomorphism.

If p does not divide n and $n \neq 1$, we have the Adem relation

$$\mathcal{P}(n) = \frac{1}{n} \mathcal{P}(1) \mathcal{P}(n-1) \ .$$

Thus, with this relation, we have an operation Γ_n defined on $(2n-1)$-dimensional classes which are annihilated by $\mathcal{P}(n-1)$ and taking values in the quotient

$$H^{2np-2}(\ \ ; Z/p)/im\mathcal{P}(1)\ .$$

We develop results about Γ_n through a general property we call a *zero theorem.* It is an analogue for Γ_n of the zero property of primary operations. We need two numerical invariants of a map, both of which are defined in terms of the Milnor filtration.

Definition 5.5.8. Given $\epsilon : X \to C$, we set

$$\text{cat } \epsilon = \min\{k | \epsilon \text{ factors through } j_{k,\infty}\}$$

where $j_{k,\infty} : B_k C \to C$ is the map from Prop. 3.5.1(c). We set cat $\epsilon = \infty$ if there is no such factorization. Given $\theta : C \to C'$, we set

$$\ell(\theta) = \min\{k | j_{k,\infty} \circ \theta \text{ is essential }\}.$$

We set $\ell(\theta) = \infty$ if there is no such minimum.

Remarks. Ganea's work [**29**] yields (for normalized category)

$$\text{cat } \epsilon \leq \ \text{cat } X\ .$$

Note that $\ell(\theta) = k$ means there is a factorization

$$\begin{array}{ccc} B_k C & \longrightarrow & (\Sigma\Omega C)^{(k)} \\ {\scriptstyle j_{k,\infty}}\downarrow & & \downarrow{\scriptstyle u} \\ C & \xrightarrow[\theta]{} & C' \end{array} \ .$$

In particular,

$$\mathcal{P}(n) : K(2n) \to K(2np)$$

has

$$\ell(\mathcal{P}(n)) = p\ .$$

We call $\ell(\theta)$ the *level* of θ. This invariant is also studied in [**97**].

5.5.1. Zero property. *Let Θ be a secondary operation based on*

$$C_0 \xrightarrow{\theta} C_1 \xrightarrow{\varphi} C_2$$

where C_0, C_1, C_2 are finite products of Eilenberg-Mac Lane spaces with finitely generated homotopy, and the homotopy groups of C_1, C_2 are Z/p-vector spaces. Suppose

$$\epsilon : X \to C_0$$

satisfies cat $\epsilon + 2 \leq \ell(\theta)$. Then ϵ is in $S_\theta(X)$ and there is a choice of tethering such that $\Theta(\epsilon) = 0$. Alternatively, any Θ based on $\varphi\theta = 0$ satisfies $\Theta(\epsilon) \equiv 0$ modulo primary operations.

Proof. Let $k = \ell(\theta)$. We prove the result for the universal example

$$j_{k-2,\infty} : B_{k-2}C_0 \to C_0 \ .$$

We have the following diagram

$$\begin{array}{ccccccc}
B_{k-2}C_0 & \longleftarrow & X & & & & \Sigma B_{k-2}C_0 \\
\downarrow & & \downarrow & & & & \downarrow{\scriptstyle \Sigma j_{k-2,1}} \\
B_{k-1}C_0 & \longrightarrow & B_kC_0 & \longrightarrow & (\Sigma\Omega C_0)^{(k)} & \longrightarrow & \Sigma B_{k-1}C_0 \\
\overline{j_{k-1,\infty}} \ \vdots & & \downarrow & & \downarrow{\scriptstyle u} & & \vdots\ {\scriptstyle \hat{u}} \\
W_\theta & \longrightarrow & C_0 & \underset{\theta}{\longrightarrow} & C_1 & \underset{\varphi}{\longrightarrow} & C_2 \\
{\scriptstyle \tilde{\varphi}}\downarrow & & & & & & \\
\Omega C_2 & & & & & &
\end{array}$$

where $\epsilon : X \to B_kC_0 \to C_0$ is factored through $B_{k-2}C_0$ by hypothesis, and u is a factorization of θ. The outer squares fill in with the dashed arrows so that the vertical composition on the left is adjoint, up to sign, to

$$\hat{u} \circ \Sigma j_{k-2,1} \ .$$

We now invoke the result, proved by Rothenberg and Steenrod [**96**], that the spectral sequence associated to this filtration on C_0 collapses at E_2. Here (for the record),

$$E_2 = \mathrm{Ext}_{H_*(\Omega C_0)}(Z/p,\ Z/p)$$

and the spectral sequence abuts at $H^*(C_0)$. It follows that the composition

$$\hat{u} \circ \Sigma j_{k-2,1} \sim w \circ \Sigma j_{k-2,2} \text{ for some } w : \Sigma B_kC_0 \to C_2 \ .$$

We use the fact $E^2 = E^\infty$ on elements giving zero in E^2 to continue the factorization, ultimately obtaining

$$\hat{u} \circ \Sigma j_{k-2,1} \sim w \circ \Sigma j_{k-2,\infty} \text{ for some } w : \Sigma C_0 \to C_2 \ .$$

This establishes the zero property. □

Question. What can be said about $\Theta(j_{k-1,\infty})$? The equation

$$\varphi \circ u = d^1 v$$

where $v : \Sigma^k \Omega C_0^{(k-1)} \to C_2$ is a factorization of $\hat{u}$, is an example of a "zig-zag" equation [**44**], [**41**]. Taken together, these equations have uses, but virtually nothing is known about the intermediate terms.

We apply the zero property to

$$\Omega\Gamma_n$$

where p does not divide n, to show that w_n is not in the image of the double suspension map

$$E^2 : S^{2n-3} \to \Omega^2 S^{2n-1} .$$

For, if it were, the mapping cone of w_n has the homotopy type of the double suspension of a complex of the form

$$X = S^{2n-3} \cup e^{2np-4} .$$

Let

$$\epsilon : X \to K(2n-3)$$

represent a generator. Then by the zero property

$$\Omega\Gamma_n(\epsilon^\flat) = 0 .$$

But this contradicts Prop. 5.5.7.

Thus far we are crossing familiar territory in an unfamiliar way. In effect, we have determined that the Toda-Hopf invariant of w_n is non-zero if p does not divide n. Moreover, the James splitting for $\Sigma J_{p-1} S^{2n-2}$ yields a homotopy equivalence

$$\Sigma(J_{p-1} S^{2n-2} \cup_{w_n^\natural} e^{2np-3}) \simeq S^{2n-1} \cup_{w_n} e^{2np-2} \vee S^{4n-3} \vee \cdots \vee S^{(2n-1)(p-1)+1} .$$

Thus, with the help of some additional cells (to get p in number) we can desuspend a complex with w_n as its only non-trivial attaching map. Now we make a new application of the zero property.

In [**12**] we introduced spaces defined as push-outs, $p \nmid n, n \geq 3$,

$$\begin{array}{ccc} S^{2n-1} & \longrightarrow & S^{2n-1} \cup_{w_n} e^{2np-2} \\ \downarrow & & \downarrow \\ S^{2n-1} \cup_{\alpha_1(2n-1)} e^{2n+2p-3} & \longrightarrow & Y_n \end{array} .$$

Proposition 5.5.9. If $p \geq 5$, the space Y_n is not homotopy equivalent to any suspension.

Proof. A putative desuspension of Y_n can be taken with cell structure

$$W = S^{2n-2} \cup e^{2n+2p-4} \cup e^{2np-2} .$$

Let $\epsilon : W \to K(2n-2)$ represent a generator. We have

$$\text{cat } \epsilon \leq \text{ cat } W \leq 3 .$$

But $\mathcal{P}(n-1)$ has level $p \geq 5$, thus

$$\Omega\Gamma_n(\epsilon) = 0 \text{ by the zero property} .$$

However, $\Gamma_n(\epsilon^\flat) \neq 0$ by Prop. 5.5.7, and naturality. □

Remark. In [**12**] it was proved that certain Y_n are cogroups. The main result in [**40**] can be used to prove that all the Y_n are cogroups. In the case $p = 3$, Ganea's desuspension theorem [**30**] yields that these Y_n have the homotopy type (even as cogroups) of suspensions. In this case, the zero property yields the information that any space W such that

$$\Sigma W \simeq Y_n \qquad (n \geq 3, 3 \nmid n)$$

has cat $W \geq 2$. Since the middle cell for the minimal W,

$$W \simeq S^{2n-2} \cup e^{2n+2} \cup e^{6n-3} ,$$

is attached by a suspension (for dimensional reasons) it follows that

$$\text{cat } W = 2 .$$

The Hopf Invariant

In the first section of this chapter, we discuss the classical Hopf invariant and related matters. The exposition is based on suggestions from J. Moore. The second section gives proofs of the Hopf invariant one theorem due, for $p = 2$, to J. F. Adams [**2**] and for p odd, independently to Liulevicius, Shimada, and Yamanoshita [**60**], [**102**].

6.1. The classical Hopf invariant

The issue in this section is a homomorphism

$$H : \pi_{2n+1}S^{n+1} \longrightarrow Z$$

from the displayed homotopy group to the integers and $n \geq 1$. We begin by observing that

$$H_*(\Omega S^{n+1}; Z) \cong T(u_n) \ ,$$

the tensor algebra with a canonical choice of generator in degree n, given by

$$u_n = h(\iota_n)$$

where h is the Hurewicz map and ι_n is the adjoint of the identity map of the $(n+1)$-sphere. Given an element

$$\alpha \text{ in } \pi_{2n+1}S^{n+1}$$

we have

$$h(\alpha^\natural) = \lambda u_n^2$$

for some integer λ, where $\alpha^\natural$ is the adjoint of α. Since λ depends only on the homotopy class, we define H by the equation

$$H(\alpha) = \lambda \ .$$

The adjoint relation is an isomorphism of groups and h is a homomorphism, thus H is also a homomorphism.

Proposition 6.1.1. If $H(\alpha) = \pm 1$, then S^n is an H-.

Proof. We extend $\alpha^\natural$ to an H-map

$$\hat{\alpha} : \Omega S^{2n+1} \longrightarrow \Omega S^{n+1}$$

by

$$\hat{\alpha} = \Omega(\text{eval.}) \circ \Omega\Sigma(\alpha^\natural) .$$

Since $\hat{\alpha}$ is an H-map, the following composition induces an isomorphism in homology

$$\Omega S^{2n+1} \times S^n \xrightarrow{\hat{\alpha}\times(\text{inc})} \Omega S^{n+1} \times \Omega S^{n+1} \xrightarrow{\text{mult}} \Omega S^{n+1}$$

and hence is a homotopy equivalence. Thus, as a retract of an H-space, S^n is an H-space. □

Proposition 6.1.2. **(a):** If n is even, H is the 0 map.

(b): If α is a suspension, $H(\alpha) = 0$.

(c): If d is a self-map of S^{n+1} of degree d, then $H(d \circ \alpha) = d^2 H(\alpha)$, note d is a map of the target.

Proof. For (a) we write, with integer coefficients suppressed,

$$H_*(\Omega S^{n+1}) = T(a) .$$

Then

$$(a \otimes 1 + 1 \otimes a)^2 = a^2 \otimes 1 + a \otimes a + (-1)^{n^2} a \otimes a + 1 \otimes a^2 .$$

Now $h(\alpha^\natural)$ is a primitive element in $T(u_n)$. The middle terms do not cancel if n is even, and the only primitive element in dimension $2n$ is 0.

For (b) we observe that the adjoint of α factors through S^n, thus $H(\alpha) = 0$ for dimensional reasons.

For (c) we observe that

$$(\Omega d)_* u_n^2 = d^2 u_n^2$$

and (c) follows. □

For the remainder of this section, n is assumed to be odd, unless otherwise stipulated. The next result is the principal device for producing maps with non-zero Hopf invariant. Given

$$f : S^n \times S^n \longrightarrow S^n$$

we say f has *bidegree* (r, s) provided that its restrictions to the axial spheres

$$f|S^n \vee S^n$$

have degree r on the first sphere and s on the second.

We construct maps by means of the following general construction. Starting with pointed CW complexes, up to homotopy, and a pointed map

$$g : X \times Y \to \Omega Z$$

we receive

$$\bar{g} : X \times Y \to \Omega Z$$

by the formula

$$\bar{g}(x, y) = g(x, *) \cdot g(*, y)$$

where $*$ is the base point in its respective factor, and $\cdot$ denotes a loop multiplication. Then we define the map

$$\alpha(g) : X \wedge Y \to \Omega Z$$

by the formula

$$\alpha(g)(x, y) = \bar{g}(x, y) \cdot \tau \circ g(x, y)$$

where τ is the direction reversal map. The homotopy extension property for the pair

$$(X \times Y, X \vee Y)$$

yields that $\alpha(g)$ is well-defined up to homotopy.

Proposition 6.1.3. Given $f : S^n \times S^n \to S^n$ with bidegree (r, s), there is an element

$$\alpha(g) : S^{2n+1} \to S^{n+1}$$

with Hopf invariant rs, where g is the composition of the inclusion map of S^n in ΩS^{n+1} with f. In particular, if S^n is an H-space, there is an element of Hopf invariant 1 in $\pi_{2n+1} S^{n+1}$.

Proof. We make a direct calculation. We abbreviate S^n to S and the k-fold product of S^n to S_k. Then, from the definition, we can factor $\alpha(g)$ as the composition

$$S_2 \xrightarrow{\Delta \times \Delta} S_4 \xrightarrow{1 \times T \times 1} S_4 \xrightarrow{\bar{g} \times \tau \circ g} (\Omega S^{n+1})_2 \longrightarrow \Omega S^{n+1} .$$

Before writing a calculation, we recall that $\tau \circ g$ induces multiplication by -1 on generators and

$$(\tau \circ g)_* = 0 \text{ on } H_{2n}(S_2)$$

because it factors through S. Then

$$\begin{aligned}
a \otimes b &\to (a \otimes 1 \otimes 1 \otimes 1 + 1 \otimes a \otimes 1 \otimes 1)(1 \otimes 1 \otimes b \otimes 1 + 1 \otimes 1 \otimes 1 \otimes b) \\
&= a \otimes 1 \otimes b \otimes 1 + a \otimes 1 \otimes 1 \otimes b + 1 \otimes a \otimes b \otimes 1 + 1 \otimes a \otimes 1 \otimes b \\
&\qquad \text{by } (\Delta \times \Delta)_* , \\
&\to a \otimes b \otimes 1 \otimes 1 + a \otimes 1 \otimes 1 \otimes b + (-1)^{n^2} 1 \otimes b \otimes a \otimes 1 + 1 \otimes 1 \otimes a \otimes b \\
&\qquad \text{by } (1 \times T \times 1)_* , \\
&\to rsu_n^2 - rsu_n^2 + (-1)^{n^2}(-1) rsu_n^2 + 0 \\
&= rsu_n^2 \text{ since } n \text{ is odd.}
\end{aligned}$$

□

The connection between the Hopf invariant and classical algebra is supplied by

Proposition 6.1.4. If $\mathbb{R}^{n+1}$ has the structure of a real division algebra, then S^n is an H-space.

Proof. A map $S^n \times S^n \to S^n$ of bidegree (1,1) comes from embedding S^n in $\mathbb{R}^{n+1}$ as the set of unit vectors. Since the multiplication has a unit and no zero divisors, we have the composition

$$S^n \times S^n \to (\mathbb{R}^{n+1} - \{0\}) \times (\mathbb{R}^{n+1} - \{0\}) \to (R^{n+1} - \{0\}) \xrightarrow{\simeq} S^n$$

where the penultimate map is the putative multiplication on $\mathbb{R}^{n+1}$ and the final map is the standard deformation retraction. □

To state the next result, we shift notation slightly to write

$$\alpha : S^{4n-1} \longrightarrow S^{2n} , \ n \geq 1 .$$

We shall describe the integral cohomology ring for the mapping cone of α, T_α, in terms of $H(\alpha)$.

Proposition 6.1.5. There are generators for $H^*(T_\alpha; Z)$,

$$x_{2n} \text{ in } H^{2n}(T_\alpha; Z) ,$$
$$y_{4n} \text{ in } H^{4n}(T_\alpha; Z) ,$$

with $x_{2n}^2 = H(\alpha) y_{4n}$.

Proof. Let $K(m)$ denote $K(Z, m)$ and

$$b_m \text{ in } H^m(K(m); Z)$$

denote the fundamental class. For dimensional reasons, we have a result like Prop. 3.5.3(a) for b_{2n}^2, there is a homotopy commutative diagram

$$\begin{array}{ccc} B_2 & \xrightarrow{\;q\;} & \Sigma K(2n-1))^{(2)} \\ {\scriptstyle j_{2,\infty}}\downarrow & & \downarrow{\scriptstyle \zeta} \\ K(2n) & \xrightarrow[b_{2n}^2]{} & K(4n) \end{array}$$

with $\zeta = \pm[b_{2n-1} \mid b_{2n-1}]$. This follows because

$$j_{1,\infty}^* b_{2n}^2 = 0$$

since B_1 is a suspension, and for $k \geq 2$,

$$j_{K,\infty}^* b_{2n}^2 \neq 0$$

since the connectivity of $(\Sigma K(2n-1))^{(k+1)}$ is greater than $4n$. The target of q has infinite cyclic integral cohomology in dimension $4n$, so ζ is some integral multiple of a generator. The same argument can be made mod p, for all primes p, so the coefficient is ± 1. We need not refine it further.

Now we let W denote the homotopy fiber of the map b_{2n}^2 and set up the diagram

$$\begin{array}{ccccccc} S^{4n-1} & \xrightarrow{\;\alpha\;} & S^{2n} & \xrightarrow{\;j\;} & T_\alpha & \longrightarrow & S^{4n} \\ & & & \searrow^{\epsilon} & & & \\ K(4n-1) & \xrightarrow[j]{} & W & \xrightarrow[p]{} & K(2n) & \xrightarrow[b_{2n}^2]{} & K(4n) \end{array}$$

with ϵ representing a generator. The lift $\bar{\epsilon}$ of ϵ is unique up to homotopy, and taking adjoints of the resulting square on the left produces a homotopy commutative diagram

$$\begin{array}{ccc} S^{4n-2} & \xrightarrow{\;\alpha^\natural\;} & \Omega S^{2n} \\ {\scriptstyle \Omega\epsilon'}\downarrow & & \downarrow{\scriptstyle \Omega\bar{\epsilon}} \\ K(4n-2) & \xrightarrow[\Omega j]{} & \Omega W \\ \| & & \downarrow{\scriptstyle \gamma^{\natural\natural}} \\ K(4n-2) & = & K(4n-2) \end{array}$$

where γ is the element from Prop. 3.5.4 corresponding to ζ and ϵ' is the factorization of $\bar{\epsilon} \circ \alpha$ through j. It follows that the degree of $\Omega\epsilon'$ is $H(\alpha)$, and we have Prop. 6.1.5 after a change of sign by invoking Cor. 3.4.3. □

Next, we exhibit elements with Hopf invariant ± 2. Let

$$w : S^{4n-1} \to S^{2n}$$

be an attaching map for the $4n$-cell in a cell decomposition for ΩS^{2n+1}, with one cell for each non-zero homology group. Let

$$f : T_w \to \Omega S^{2n+1}$$

be a map from the mapping cone of w which extends the inclusion of the bottom cell (adjoint of the identity map of S^{n+1}). Then f induces an isomorphism of integral homology in dimensions $\leq 6n - 1$. We may use this isomorphism to define the following elements, (subscripts denote degree)

$$c_{2n} \ , \ c_{4n} \text{ in } H^*(\Omega S^{2n+1}; Z)$$

satisfy

$$\langle u_{2n}, \ c_{2n} \rangle = 1 = \langle u_{2n}^2, \ c_{4n} \rangle$$

under the Kronecker pairing. We obtain generators for the integral cohomology of T_w,

$$x_{2n} = f^* c_{2n}, \ y_{4n} = f^* c_{4n} \ .$$

Now the calculation in Prop. 6.1.2(a) yields

$$\langle u_{2n}^2, \ c_{2n}^2 \rangle = 2 \ ,$$

hence

$$c_{2n}^2 = 2c_{4n}$$

since the groups are infinite cyclic. It follows by naturality that

$$x_{2n}^2 = f^* c_{2n}^2 = 2f^* c_{4n} = 2y_{4n} \ ,$$

hence

$$H(w) = \pm 2 \ .$$

The sign ambiguity arises because the generator denoted by y_{4n} here may differ by sign from that employed in Prop. 6.1.5. □

In the sequel it will be advantageous to note that having elements with Hopf invariant 2 in hand implies that there exists an element of Hopf invariant one if and only if there exists an element with odd Hopf invariant.

We proceed to treat in concert, some further results about the first attaching maps in the complexes:

(1): $\Omega S^{n+1} \simeq S^n \cup_{w_n} e^{2n} \cup \cdots \ ,$

(2): $\Omega J_{p-1} S^{2n} \simeq S^{2n-1} \cup_{w_n} e^{2np-2} \cup \cdots \ ,$

and we have written w_n for this map. For (1), 2 is the prime of interest, while for (2), p is an odd prime. In the case of (1), w_n may be identified with the Whitehead product. We shall not make use of that identification here, but instead treat these maps using the information gained as attaching maps. The reader is referred to [**25**] and [**99**] for more developments beyond those pursued here.

Our discussion will be carried out in terms of pairs of spaces and maps between them,

$$v : A \to \Omega X ,$$

which exist under various hypotheses on w_n. There will also be two critical dimensions, d and c. All this data will be associated with the complexes in (1) or (2). We now list the data. The Moore space obtained by attaching an n-cell to S^{n-1} by a map of degree x is written $P^n(x)$.

In case (1), $d = n+1$, $c = 2n$, $X = S^d$; and under the hypothesis $w_n = 0$, A is S^c, while under the hypothesis that 2 divides w_n, then $A = P^c(2)$.

In case (2), $d = 2n$, $c = 2np - 2$, $X = J_{p-1}S^d$; and under the hypothesis $w_n = 0$, $A = S^c$ while under the hypothesis that p divides w_n, then $A = P^c(p)$.

In these terms, we can construct a map

$$v : A \to \Omega X$$

with non-zero Hurewicz image. For example, we receive

$$v : P^{2np-2}(p) \to \Omega J_{p-1} S^{2n}$$

with non-zero Hurewicz image, if p divides w_n, from the following diagram:

$$\begin{array}{ccc}
S^{2np-3} & \xrightarrow{=} & S^{2np-3} \\
\downarrow{\scriptstyle \times p} & & \downarrow{\scriptstyle w_n} \\
S^{2np-3} & \xrightarrow{w'} & S^{2n-1} \\
\downarrow & & \downarrow \\
P^{2np-2}(p) & \xrightarrow[v]{} & \Omega J_{p-1} S^{2n} .
\end{array}$$

Now we look at $T =$ mapping cone of the adjoint of v

$$v^\flat : \Sigma A \to X \xrightarrow{j} T .$$

Proposition 6.1.6. Let x in $H^d(T; Z/p)$ be a generator. Then $x^p \neq 0$.

Proof. In each case, we chase in the diagram for the cohomology Serre spectral sequences for the path-loop fibration over X and T with the map between them induced by j. Moreover, the argument is indirect, assume

$$x^p = 0 .$$

For the case detailed above, we have (Z/p coefficients are suppressed)

$$x = d_{2n}y$$

for some y in $H^{2n-1}(\Omega T)$. If $x^p = 0$, then there is

$$z \text{ in } H^{2np-2}(\Omega T)$$

such that

$$d_r z = x^{p-1} \otimes y\ ,\ r = 2n(p-1) - 1$$

since this is the first and last possible differential to eliminate the "product" term. By the same chasing as done in Chapter 5, section 5 it follows that in cohomology

$$(\Omega j)^* \text{ is surjective in dimension } 2np - 2\ .$$

Hence, by duality in homology

$$(\Omega j)_* \text{ is injective in dimension } 2np - 2\ .$$

But this means that, in homology in this dimension that

$$(\Omega j \circ v)_* \text{ is injective}$$

because v has non-zero Hurewicz image. But, by adjointness $\Omega j \circ v$ factors through

$$\Omega j \circ \Omega v^\flat$$

which is null-homotopic, by construction.

We leave the other cases for the reader. I must note that I learned this argument from Selick's paper [**99**]. □

Let us now look at some consequences of Prop. 6.1.6. In the situation of (1) and $w_n = 0$, we have, an element of Hopf invariant one and a space,

$$T = S^{n+1} \cup e^{2n+2}$$

with non-trivial action of $Sq(n+1)$. Hence $(n+1)$ must be a power of 2, by the Adem relation.

In the situation of (1) with w_n divisible by 2 we have,

$$T = S^{n+1} \cup e^{2n+1} \cup e^{2n+2}$$

with $Sq(n+1)x = Sq(1)y$ for some y.

It is easy to check that the Adem relations force $(n+1)$ to be a power of 2, or $n = 2$.

In the situation of (2) with $w_n = 0$, we have

$$T = J_{p-1}S^{2n} \cup e^{2np}\ .$$

We can use James splitting for $\Sigma J_{p-1}S^{2n}$ to pinch off the intermediate spheres with the result that we have a space of the form

$$S^{2n+1} \cup e^{2np+1}$$

with a non-trivial $\mathcal{P}(n)$ action. Thus n must be a power of p by the Adem relations.

In the situation of (2) with w_n divisible by p, we have

$$T = J_{p-1}S^{2n} \cup CP^{2np-1}(p)$$

and we can again suspend T and pinch off intermediate spheres to obtain a complex of the form

$$S^{2n+1} \cup e^{2np} \cup e^{2np+1}$$

with $\mathcal{P}(n)x = \beta y$. By the Adem relations, n must be a power of p.

The overall conclusion is that, except in the prime power case, w_n is non-zero and not divisible by p. The prime power case is the story for the next section. Meanwhile, we draw one further conclusion from our arguments.

Proposition 6.1.7. The inclusion map $S^4 \to HP^\infty$ does not kill w_4, the attaching map of the 8-cell to S^4 in ΩS^5.

Proof. Through the 8-skeleton, HP^∞ is

$$S^4 \cup_v e^8$$

where $\nu = v_4$, the element with Hopf invariant one. If $w_4 = 0$ in HP^∞, we have a map

$$f : S^4 \cup_w e^8 \to S^4 \cup_v e^8$$

extending the identity map on S^4. Calculating in integral cohomology and using Prop. 6.1.5 together with the fact that the Hopf invariant of w is ± 2 yields that f^* has degree ± 2 in dimension 8. We feed this information into the ladder of homotopy exact sequences induced by f,

$$\begin{array}{ccccc} \pi_8(S^4 \cup_w e^8, S^4) & \xrightarrow{\partial} & \pi_7 S^4 & \longrightarrow & \pi_7(S^4 \cup_w e^8) \\ {\scriptstyle \text{mult. by } \pm 2}\downarrow & & \| & & \downarrow \\ \pi_8(HP^2, S^4) & \xrightarrow{\partial} & \pi_7 S^4 & \longrightarrow & \pi_7(HP^2) \end{array}$$

using the Hurewicz Theorem on the left, to conclude that $w = \pm 2v$. But the homotopy sequence for the pair

$$(\Omega S^5, S^4)$$

shows that w is killed by suspension. On the other hand, we know that $2v$ is detected by the Adams operation $\Phi_{0,2}$ which is stable. $\square$

A similar argument shows that the inclusion of S^8 into the Cayley projective plane

$$W = S^8 \cup_\sigma e^{16}$$

does not kill w_8. Both results stand in contrast to the fact that

$$w_2 = 2\eta_2 \text{ in } \pi_3 S^2 = Z\ ,$$

because here, the Hopf invariant

$$H : \pi_3 S^2 \to Z$$

is an isomorphism, since η_2 has Hopf invariant one.

6.2. Hopf invariant one

Unless otherwise stipulated, the integer n is a prime power. Can there be a cell complex with only 2 cells in positive dimensions (a two cell complex) and carrying a non-trivial action of $\mathcal{P}(n)$ or $Sq(n)$? For $p = 2$ and $n = 1, 2, 4, 8$ or p odd and $n = 1$, we have examples. The celebrated Hopf invariant one theorems assert that these are the only possibilities. We have already traced the relation between this problem and the classical Hopf invariant, for $p = 2$. The introduction given by Adams [**2**] has a full account of other connections. The introduction by Liulevicius [**60**] describes connections in addition to those with the equation $w_n = 0$. We say no more about it here.

The existence of such a complex implies the existence of a stable map between spheres in the $(n-1)$-stem if $p = 2$, and the $(2n(p-1)-1)$-stem if p is odd. The term *stem* means the difference in dimension, source minus target. The map is *stable* (essential under iterated suspensions) because the primary operations are stable. The existence of the Adams spectral sequence implies that if such a complex is impossible, then $Sq(n)$ or $\mathcal{P}(n)$ must be decomposible (*a la* the Adem relations) but by means of higher order operations. Calculation of the E_2 term supplies many candidates, so perhaps it is surprising that secondary operations will suffice for the remaining values of n. Of course, if we can produce directly a relation among secondary operations and show that it works, then we have no logical need for the Adams spectral sequence (indeed, it is not mentioned in [**2**]).

The author (along with most homotopy theorists of his generation) has calculated initial segments of minimal resolutions by hand. Here are his results for the first remaining cases, $n = 16$, $n = p$. For the prime 2, the notation is from the table in (5.3.2). A correct calculation is not unique.

Write

$$A = \begin{bmatrix} 1 \\ 2 \\ 4 \\ 8 \end{bmatrix}, \quad \begin{bmatrix} \beta \\ \mathcal{P}(1) \end{bmatrix},$$

$$B = \begin{bmatrix} 1 & 0 & 0 & 0 \\ 3 & 2 & 0 & 0 \\ 4 & 2.1 & 1 & 0 \\ 7 & 6 & 4 & 0 \\ 8 & 7 & 4.1 & 1 \\ 7.2 & 8 & 4.2 & 2 \\ 15 & 14 & 12 & 8 \end{bmatrix}, \quad \begin{bmatrix} 2\mathcal{P}(2)\,, & \beta\mathcal{P}(1) - 2\mathcal{P}(1)\beta \\ 0\,, & \mathcal{P}(p-1) \end{bmatrix},$$

$$C = [15 + 11.4,\ 11.2,\ 12 + 11.1,\ 8.1 + 6.3 + 6.2.1,\ 8 + 6.2,\ 7 + 4.2.1,\ 1],$$
$$[\mathcal{P}(p-2),\ -\beta]\,.$$

Please check that the Adem relations yield

$$BA = 0,\ CB = 0\,.$$

From this data, we can construct syzygys of length 3.

Write

$$K(m+:\ i_1, \ldots, i_k) \text{ for the product}$$

$$\mathop{\mathrm{X}}_{j=1}^{k} K(Z/p,\ m + i_j)$$

For the primes 2 and odd respectively, write

$$C_1 = K(m+ : 1, 2, 4, 8), \quad K(m+ : 1, q) \text{ where } q = 2p - 2\,,$$
$$C_2 = K(m+ : 2, 4, 5, 8, 9, 10, 16), \quad K(m+ : 2, 2q + 1, pq)\,,$$
$$C_3 = K(m + 17), \qquad K(m + pq + 1)\,.$$

Then we have maps

$$A_m : K(m) \to C_1\,,$$
$$B_m : C_1 \to C_2\,,$$
$$C_m : C_2 \to C_3\,,$$

where the signed Bockstein $(-1)^m\beta$ has been inserted for the odd prime case. We have stable secondary operations

$$\Theta_m : S_{A_m}(\quad) \to T_{\Omega B_m}(\quad)$$

based on $[B_m][\tau_m \circ A_m] = 0$. Moreover, the differences in tetherings are measured by

$$[K(m), \Omega C_2]\,,$$

which is in the image of $A_m^{\#}$ because the Steenrod algebra is generated through dimensions 15 and $pq - 1$ by the entries of A_m, and we may invoke the strong invariance property Prop. 3.3.5 for coliftings constructed for small values of m.

The components of Θ_m are the Adams and $L{-}S{-}Y$ operations discussed in Chapter 5. In particular, for $p = 2$, the components are,

$$\Phi_{i,j},\ 0 \le i \le j \le 3,\ j \ne i+1\,,$$

and for odd primes,

$\mathcal{R}_m$, and an operation associated with λ_0, which Liulevicius writes as

$$\Gamma_{\lambda_0}$$

(The reader of [**60**] will discover that the secondary operations of odd degree are indexed by their name in the cohomology of the Steenrod algebra.)

There are several ways to process the information latent in our set-up. They are represented in the following diagram and the processing is by means of the manipulations set out under subheading (4.2.5).

$$\begin{array}{ccccc}
W_m & \xrightarrow[\boxed{-1}]{\tilde{B}_m} & \Omega C_2 & \xrightarrow{\Omega C_m} & \Omega C_3 \\
\downarrow & & \downarrow & & \| \\
K(m) & \xrightarrow{\alpha} & I_m & \xrightarrow{\tilde{C}_m} & \Omega C_3 \\
\| & & \downarrow & & \\
K(m) & \xrightarrow[\tau_m \circ A_m]{} & C_1 & \xrightarrow[B_m]{} & C_2
\end{array}$$

where $W_m = W_{\tau_m \circ A_m}$ and $I_m = W_{B_m}$ are the homotopy fibers of the subscripted maps, and

$$\alpha = \overline{\tau_m \circ A_m}$$

is a lifting. The anti-commutative part, indicated by the boxed unit, is from subheading (3.2.7). Moreover,

$$\Omega\tilde{B}_m \text{ and } \widetilde{B_{m-1}}$$

are compatible, in the sense that

$$\Omega\Theta_m = \Theta_{m-1}$$

but

$$\Omega\tilde{C}_m \text{ and } \widetilde{C_{m-1}}$$

are off by a sign, because $\tilde{C}_m$ is a colifting associated with a null-homotopy for

$$C_m \circ B_m$$

without incorporating the map τ_m. The author is fussing because he has drawn mistaken conclusions from arguments depending on cancellation.

Most importantly, the value of

$$\Omega C_m \circ \tilde{B}_m$$

is independent of tetherings for Θ_m and

$$\tilde{C}_m \circ \alpha$$

represents this value, up to sign in

$$[K(m), \Omega C_3] \Big/ _{im\ A_m^{\#} + im\ \Omega C_{m\#}} ,$$

and the representative is a loop class. A peek at the dimensions for C_3 reveals that our representative is some multiple of

$$Sq(16)b_m,\ \mathcal{P}(p)b_m$$

and the question is whether the coefficient is 0 or not.

It is not 0. To see this, we map CP^∞ to $K(m)$ for $m = 16,\ 2p$ by

$$CP^\infty \xrightarrow{\epsilon} K(16) \quad \text{represents } u^8 ,$$

$$CP^\infty \xrightarrow{\epsilon} K(2p) \quad \text{represents } u^p .$$

Now Θ_m is defined on ϵ, as is easily checked. For dimensional reasons, the only Adams operations

$$\Phi_{0,2},\ \Phi_{0,3}$$

and the Liulevicus operation

$$\mathcal{R}_{2p}$$

can yield a non-zero value. We have made these calculations in Prop. 5.3.5 and Prop. 5.4.1. The results respectively are

$$0,\ [\![u^{12}]\!],\ [\![u^{3p-2}]\!]$$

with 0 indeterminacy. Consequently, the only components of ΩC_m which can affect the outcome are

$$Sq(8) + Sq(6,2) \text{ and } \mathcal{P}(p-2) .$$

Now

$$(Sq(8) + Sq(6,2))u^{12} = u^{16} = Sq(16)u^8$$

and

$$\mathcal{P}(p-2)(u^{3p-2}) = \binom{3p-2}{p-2} u^{p^2} = \mathcal{P}(p)u^p .$$

Thus the coefficient is non-zero in both cases. □

We summarize in

Theorem 6.2.1. **(a):** Let x be an m-dimensional mod 2 cohomology class which is annihilated by $Sq(1), Sq(2), Sq(4), Sq(8)$. Then

$$\begin{aligned} Sq(16)x = {} & (Sq(15) + Sq(11,4))\Phi_{0,0}(x) + Sq(11,2)\Phi_{1,1}(x) \\ & + (Sq(12) + Sq(11,1))\Phi_{0,2}(x) + (Sq(8,1) + Sq(6,3) \\ & + Sq(6,2,1))\Phi_{2,2}(x) \\ & + (Sq(8) + Sq(6,2))\Phi_{0,3}(x) + (Sq(7) + Sq(4,2,1))\Phi_{1,3}(x) \\ & + Sq(1)\Phi_{3,3}(x) \text{ modulo the total indeterminacy,} \end{aligned}$$

i.e. in

$$H^{m+16}(X; Z/2) \Big/ \sum_{(i,j)} \text{Ind } (\alpha_{i,j}\Phi_{i,j}, X)$$

where $\alpha_{i,j}$ is the displayed coefficient of $\Phi_{i,j}$.

(b): Let x be an m-dimensional mod p cohomology class which is annihilated by the Bockstein and $\mathcal{P}(1)$. Then

$$\mathcal{P}(p)x = \mathcal{P}(p-2)\mathcal{R}(x) - \beta\Gamma_{\lambda_0}(x)$$

in

$$H^{m+2p(p-1)}(X; Z/p) \Big/ \beta\mathcal{P}(p-1)H^{m+2p-3} + \mathcal{P}(p-1)\beta H^{m+2p-3} .$$

Before passing to the remaining prime power cases, we use (a) to discuss the question raised in exercise (1.2.3). We ask whether there is a space with mod 2 cohomology ring

$$Z/2[w] \Big/ (w^4), \ |w| = 8 .$$

We set up a way to study this question, following Goncalves [**36**]. The calculation is completed in Chapter 7.

To study this situation, we extract an operation from the final three entries of the 1×7 matrix C and the final column of B,

$$\varphi = [Sq(8) + Sq(6,2), \ Sq(7) + Sq(4,2,1), \ Sq(1)]$$

and

$$\theta = \text{ col } [Sq(1), \ Sq(2), \ Sq(8)] .$$

We write $\Phi'_{0,3}$ for a stable secondary operation based on $[\varphi][\theta] = 0$. Then, reading Theorem 6.2.1 through the components of

$$\tilde{C}_m \circ \alpha$$

and employing the set-up from part (g) of subheading (4.2.5), we obtain

$$[\![Sq(16)(p^*b_m)]\!] = \Phi'_{0,3}Sq(8)(p^*b_m) \text{ up to other primary operations.}$$

The diagram is displayed below:

$$\begin{array}{ccccccc}
W_m & \xrightarrow{Sq(8)\circ p} & K(m+8) & \xrightarrow{\begin{pmatrix} Sq(1) \\ Sq(2) \\ Sq(8) \end{pmatrix}} & K(m+:9,10,16) & \xrightarrow{\varphi} & K(m+17) \\
\Big\downarrow{\scriptstyle p} & & \Big\downarrow{\scriptstyle j} & & \Big\downarrow{\scriptstyle j} & & \Big\| \\
K(m) & \xrightarrow[A_m]{} & C_1 & \xrightarrow[B_m]{} & C_2 & \xrightarrow[C_m]{} & K(m+17) \\
 & & \Big\downarrow{\scriptstyle \rho} & & & & \\
 & & K(m+:1,2,4) & & & &
\end{array}$$

where W_m is the fiber of $\rho \circ A_m$ and an unsubscripted ρ or j is the obvious projection or inclusion map and the other maps exploit the zeros in C_m. In this situation, a map from X to $K(m)$

$$X \xrightarrow{\epsilon} K(m)$$

lifts to W_m if and only if each composition

$$Sq(i) \circ \epsilon, \ i = 1, 2, 4$$

is null-homotopic. Moreover, we may regard the diagram for m as obtained by looping a diagram for $(m+1)$, as $p = 2$ and signs give no difficulty. So our equation is one of stable operations, then the ambiguous primary terms are stable. In particular, for $m = 8$, the left side gives 0. In Chapter 7 we shall establish that if w is a mod 2 cohomology class of degree 8 which is annihilated by $Sq(i)$ for $i = 1, 2, 4$, then

$$\Phi'_{0,3}Sq(8)w = [\![w^3]\!] \ .$$

This equation dismisses $Z/2[w]/(w^4)$ as a possible ring for the mod 2 cohomology of any space. Intuitively, this formula results because we can put the $Sq(1)$ against $Sq(8)$ to get $Sq(9)$ and concentrate our attention on a stable operation built from $Sq(8)Sq(9)$, on 8-dimensional classes. As in Chapter 5, section 1, we expect the value of the stable operation to involve $u_8 \cup Sq(8)u_8$ in its universal example. Besides the manipulations of (4.2.5), we shall use the structure theory developed in the next chapter to work out enough of the cohomology of W_m for a precise argument. It is possible to use the method of Chapter 5 to work this example, but this entails a number of steps that are unnecessary in light of the structure theory.

Because I've been asked, I point out that no assumption of simple connectivity will be necessary for our argument. I should also note that the other components of $\tilde{C}_m \circ \alpha$ could be considered the same way, but they give 0 for this problem.

As a counter-balance to the free use of variations in the Adem relations used to define operations, we look at another example, one illustrating an error that the author has made on occasion. We can base an operation $\mathcal{R}'$ on the relation $[\varphi][\theta] = 0$ where

$$\varphi = [2\mathcal{P}(2),\ \mathcal{P}(1),\ 2\beta]$$
$$\theta = \operatorname{col}\,[\beta, \beta\mathcal{P}(1),\ \mathcal{P}(2)]\ .$$

We may compare $\mathcal{R}$ with $\mathcal{R}'$ and find that they agree up to primary operations on their common domain. Similarly, we can base an operation Γ'_{λ_0} on

$$\mathcal{P}(p-2)\mathcal{P}(2) = 0$$

and likewise compare it with Γ_{λ_0} of Liulevicius. But it does not follow that

$$\mathcal{P}(p-2)\mathcal{R}' - \beta\Gamma'_{\lambda_0}$$

factors anything, let alone $\mathcal{P}(p)$. We cannot make any argument, because

$$(\mathcal{P}(p{-}2), -\beta)\begin{pmatrix} 2\mathcal{P}(2),\ \mathcal{P}(1),\ 2\beta \\ 0,\ 0,\ \mathcal{P}(p-2) \end{pmatrix} = (0, \mathcal{P}(p{-}2)\mathcal{P}(1), 2\mathcal{P}(p{-}2)\beta{-}\beta\mathcal{P}(p{-}2))$$

which is not 0, and fiddling with coefficients is no help. Thus we do not have a syzygy.

In fact, when $p = 3$, it is impossible to factor $\mathcal{P}(p)$ in the presence of $\mathcal{P}(1)$. In [**123**], Zabrodsky produces a space with mod 3 cohomology algebra a polynomial algebra on two generators of degree 12, 16, and

$$\mathcal{P}(1)x_{12} = x_{16}\ .$$

We have

$$\mathcal{P}(6)x_{12} = x_{12}^3 \neq 0\ .$$

The reader may check that if $\mathcal{P}(3)$ factored through $\mathcal{R}$ and Γ_{λ_0} when $p = 3$, even in variant form, there would be a contradiction.

A significant generalization, of the non-factorability of

$$\mathcal{P}(p^k)$$

in torsion free spaces, in the presence of

$$\mathcal{P}(1), \ldots, \mathcal{P}(p^{k-1})$$

is given by Kane [**49**]. His arguments are outside the scope of this book. The best result for $p = 2$ is due to Hubbuck [**46**].

As an additional caveat offered by this example, note that

$$\mathcal{P}(p-2)\mathcal{R}'$$

is non-trivial on u^p in CP^∞. A similar observation applies to

$$Sq(4)\Phi_{0,2} \text{ on } u^4 .$$

Without the syzygy, these are just isolated calculations. They illustrate a sticky technical point: secondary compositions, and their higher order variants, may be easy to prove non-zero provided they are defined. But the calculation is no guarantee of existence.

We turn to the remaining prime powers. We write

$$n = p^{N+1}$$

and look for factorizations of $Sq(n)$ and $\mathcal{P}(n)$, where for p odd, $N \geq 1$. In view of earlier work, we may assume $N \geq 4$ if $p = 2$, but without further comment, we set $N \geq 3$. In view of the results from section 3 of Chapter 5, we have

$$\lambda Sq(n) = \Sigma c_{i,j,n}\Phi_{i,j} \text{ modulo the total indeterminancy}$$

where

$$0 \leq i \leq j \leq N,\ j \neq i+1 \text{ and}$$
$$c_{0,N,n} = Sq(2^N) + \text{ decomposables}$$
$$c_{N,N,n} = Sq(1) .$$

From Prop. 5.3.5 and the fact that operations indexed by (i,j) with $i \neq 0$ have odd degree, we obtain $\lambda = 1$. Putting $\lambda = 1$ is the celebrated *Adams decomposition formula.* We have discussed how it settles the Hopf invariant one question. □

In view of the results from Chapter 5, section 4, we have

$$\lambda\mathcal{P}(n) = \sum_{k=1}^{N} a_{k,N}\Psi_k + b_N\mathcal{R} + \sum_{\gamma} c_{N,\gamma}\Gamma_\gamma$$

where we have used Liulevicius notation. We have

$$a_{N,N} = \mathcal{P}(p^N(p-1)) + \text{ other decomposable terms.}$$

Now

$$\mathcal{P}(p^N(p-1))u^{p^N(2p-1)} = u^{p^{N+2}} .$$

The Lucas formula is used to see that other possible terms give 0 on this element. In particular, write

$$I = (S_1, \ldots, S_T) \quad T \geq 2 .$$

Then

$$\binom{p^N(2p-1)}{S_T} = 0$$

unless the p-adic expansion on S_T has the form

$$C_N P^N + C_{N+1}p^{N+1} ,$$

with $0 \le C_N \le p-1$, $0 \le C_{N+1} \le 1$, and all other coefficients equal to 0. Moreover,

$$S_1 + \cdots + S_T = p^N(p-1)$$

and

$$S_{T-1} \ge pS_T\ .$$

Those are impossible equations if $T \ge 2$. Hence, we may use Proposition 5.4.2 and the comparison of Ψ_K with Ψ'_K to conclude $\lambda \neq 0$. Understanding $\lambda \neq 0$ yields the *decomposition formula* due to Liulevicius, Shimada and Yamanoshita. □

Diagrams may be set up for these factorizations, just as for Theorem 6.2.1. One consequence is a reformulation in which only primary operations are mentioned. Thus if X is a space with a mod 2 cohomology class x satisfying

$$Sq(1)x = Sq(2)x = \cdots = Sq(2^N)x = 0\ ,$$

then there are classes $y_{i,j}$ and primary operations $\alpha_{i,j}$ of positive degree such that

$$Sq(2^{N+1})x = \sum_{i,j} \alpha_{i,j} y_{i,j}\ .$$

Another example of this style for stating facts derived from secondary operations appears in subheading (5.2.2). The author calls them *conditional relations* and more appear in Chapter 7.

Finally, we indicate what this work does for the question concerning the divisibility of w_n by p. The decomposition formulas do not answer the question, but point to secondary operations which would detect homotopy if such divisibility were possible, and indicate necessary properties of the map so detected. In particular, the decomposition formulas do not rule out the complexes displayed in Section 1.

The reader may be curious as to what is known about the problem. First of all w_p is divisible by p (p odd) by direct construction. For example, when $p = 3$ we have (up to non-zero constant)

$$w_3 = \alpha_1\alpha_2 = 3 \times \beta_1$$

on the 5-sphere.

The only general result is due to Ravenel [**95**] and concerns $p \ge 5$. Using advanced, sophisticated methods Ravenel proves a result about the Adams spectral sequence. A corollary is that p does not divide w_n for $p \ge 5$ because the existence of the complex would yield (in view of the $L-S-Y$ decomposition formula) that

$$\Gamma_{\lambda_N}$$

detect homotopy and this is what Ravenel proves is impossible.

The situation for $p = 3$ is in a peculiar state of uncertainty. Toda established the famous Toda differential [**112**] which yields that 3 does not divide w_9. Actually Toda proves a result that has the consequence

$$p \text{ does not divide } w_{p^2} \,.$$

Toda's result grounds an induction for Ravenel at $p \geq 5$. The induction breaks down for $p = 3$. Ravenel reports in [**95**] that

$$\Gamma_{\lambda_2}$$

does detect homotopy (but of unknown order) in the stable homotopy of spheres, but this does not settle the divisibility issue for w_{27}.

For $p = 2$ we have that

$$w_{15},\ w_{31},\ w_{63}$$

are divisible by 2. The first is a benign exercise for advanced students of homotopy theory. The other two are due to Barratt and Mahowald [**65**], and also established by Milgram [**79**]. After that, the question is open. A good place to begin learning about this question, as well as its connection with the Kervaire-Milnor [**52**] paper is found in [**10**]. A survey of recent work is given by N. Minami, "On the Kervaire invariant problem," in Contemporary Mathematics **220**, Homotopy theory via algebraic geometry and group representations (eds. M. Mahowald, S. Priddy), Amer. Math. Soc. (1998) 229–254.

Exercise 6.2.2. Use an Adem argument with the $L - S - Y$ decomposition formula, together with Prop. 5.5.4 to prove: If $p\times$ an element detected by Γ_{λ_N} is 0, then p divides w_n where $n = p^{N+1}$. Formulate and prove a similar result for $p = 2$.

Chapter 7

The Cohomology Structure of Universal Examples

In this chapter we develop detailed information for the cohomology structure of the fiber spaces which arise in our constructions, in the case where we are working with Steenrod operations. We can then place the geometric constructions of Chapter 3 in an algebraic context and work out relations which are not so evident on elementary grounds. For matters involving the principal action or the H-space multiplication, the geometric work lays the foundation for important calculations which can be made directly, and not inferentially as is often the case in the literature. We also apply our methods, now using both algebra and geometry to a discussion of twisted operations and Cartan formulas.

7.1. Unstable modules and algebras over the Steenrod algebra

We start with the category of graded modules over the Steenrod algebra, and degree preserving maps. We write $\mathcal{U}$ for the full subcategory whose objects satisfy the zero property for the action of the Steenrod algebra.

Definition 7.1.1. An $\mathcal{A}$-module M is unstable provided that for each n,

$$B(n) \cdot M^n = 0$$

where $B(n) \subset \mathcal{A}$ is the submodule of n-annihilators. An equivalent formulation in terms of elements is:

$$\text{for } p = 2, \quad Sq^i x = 0 \text{ if deg } x < i \;,$$
$$\text{for } p \text{ odd}, \; \mathcal{P}^i x = 0 \text{ if deg } x < 2i \;,$$
$$\beta \mathcal{P}^i x = 0 \text{ if deg } x \leq 2i \;.$$

The mod p cohomology of a topological space is an unstable object. We bring in the cup product structure by means of the notion of an unstable algebra. Let $\mathcal{K}$ denote the category of unstable algebras over $\mathcal{A}$, where objects are algebras over the Steenrod algebra which are unstable $\mathcal{A}$-modules, and maps preserve degree.

7.1.1. Algebraic suspensions. We define two constructions on $\mathcal{A}$-modules by means of shift operators s, s^-;

$$(sM)^{n+1} = M^n = (s^- M)^{n-1} \;.$$

If x is an element of M, we write sx (resp. $s^- x$) for the corresponding element in sM (resp. $s^- M$). Following the standard sign convention, the $\mathcal{A}$-module structure is given by

$$\theta(sx) = (-1)^{|\theta|} s(\theta x) \;,$$

and likewise for s^-.

Clearly sM is an unstable module if M is. A characterization for $s^- M$ to be unstable is that M be connected ($M^0 = 0$) and

$$B(n-1) \cdot M^n = 0 \text{ for all } n \;.$$

We use the shift operator s to define a functor on $\mathcal{U}$ called the algebraic suspension,

$$\Sigma M = sM$$

and for $f : M \to N$, $\Sigma f : \Sigma M \to \Sigma N$ is given by $\Sigma f(sx) = sf(x)$.

We may identify $H^{*+1}(\Sigma X; Z/p)$ with $s\bar{H}^*(X; Z/p)$ by means of (either) Mayer-Vietoris suspension isomorphism. Thus the algebraic suspension corresponds to geometric suspension for (non-empty) spaces. The analogous result for looping is developed next.

7.1.2. Algebraic loop functor. We introduce a functor on $\mathcal{U}$, denoted by Φ, which concentrates degrees;

$$(\Phi M)^{2n} = M^n \text{ for } p = 2,$$
$$(\Phi M)^{2np} = M^{2n},$$
$$(\Phi M)^{2np+2} = M^{2n+1} \text{ for odd primes } p,$$

and $(\Phi M)^k = 0$ for k not of the form listed above.

We write $\bar{x}$ in ΦM for x in M. Given a map $f : M \to N$, then

$$\Phi(f) : \Phi M \to \Phi N$$

is given by $\Phi(f)(\bar{x}) = \overline{(fx)}$.

ΦM receives the structure of an $\mathcal{A}$-module via the formulas,

$$\begin{aligned} Sq^{2i}\bar{x} &= \overline{(Sq^i x)}, \\ Sq^{2i+1}\bar{x} &= 0 \qquad \text{for } p = 2 , \\ \mathcal{P}^{pj}\bar{x} &= \overline{(\mathcal{P}^j x)}, \\ \mathcal{P}^{pj+1}\bar{x} &= \overline{(\beta\mathcal{P}^j x)} \text{ if } |x| \text{ is odd}, \\ \beta^\epsilon\mathcal{P}^j\bar{x} &= 0 \quad \text{ otherwise, for } p \text{ odd and } \epsilon = 0, 1 . \end{aligned}$$

Note, there is no formula for $\overline{(\beta\mathcal{P}^j x)}$ if $|x|$ is even.

Direct checking yields that ΦM is an unstable module if M is.

We have a natural transformation

$$\lambda : \Phi \to Id$$

given by

$$\begin{aligned} \lambda\bar{x} &= Sq^n x \text{ if } |x| = n \text{ and } p = 2 , \\ \lambda\bar{x} &= \mathcal{P}^n x \text{ if } |x| = 2n , \\ \lambda\bar{x} &= \beta P^n x \text{ if } |x| = 2n + 1 \text{ and } p \text{ is odd.} \end{aligned}$$

The map λ is $\mathcal{A}$-linear and both its kernel and cokernel are suspensions in $\mathcal{U}$. With these observations, we may define two more functors on $\mathcal{U}$, by the exact sequence

$$0 \longrightarrow \Sigma\Omega_1 M \longrightarrow \Phi M \xrightarrow{\lambda} M \longrightarrow \Sigma\Omega M \longrightarrow 0 .$$

Note $\lambda : (\Phi M)^0 \cong M^0$. The functor Ω is the <u>algebraic loop functor</u>. We write its elements in terms of the equation

$$\Omega M = s^- \text{ coker } \lambda .$$

Thus for x in M we have $s^-[x]$ for the corresponding element of ΩM. For maps $f : M \to N$, we have

$$\Omega f : \Omega M \to \Omega N$$

given by $\Omega f(s^-[x]) = s^-[fx]$.

We introduce an analogue of the cohomology suspension,

$$\sigma^* : M \to \Omega M$$

given by $\sigma^*(x) = s^-[x]$.

We have the commutative diagram (in $\mathcal{U}$)

$$\begin{array}{ccccccc} \Phi H^*X & \longrightarrow & H^*X & \xrightarrow{s\sigma^*} & \Sigma\Omega H^*X & \longrightarrow & 0 \\ & & \| & & \downarrow & & \\ & & H^*(X) & \underset{p_1^*}{\longrightarrow} & H^*(\Sigma\Omega X)\ , & & \end{array}$$

but the induced map on the right is neither monic nor epic in general.

The following properties of the algebraic loop and suspension functors are readily established, with details left as an exercise.

$$\Omega\Sigma = id,$$
$$\mathrm{Hom}_{\mathcal{U}}(M, \Sigma N) = \mathrm{Hom}_{\mathcal{U}}(\Omega M, N)\ .$$

A short exact seqeunce in $\mathcal{U}$,

$$0 \to A \to B \to C \to 0$$

yields an exact sequence,

$$0 \to \Omega_1 A \to \Omega_1 B \to \Omega_1 C \to \Omega A \to \Omega B \to \Omega C \to 0$$

with the connecting homomorphism obtained via the switchback construction and the maps

$$\Sigma\Omega_1 C \to \Phi C \leftarrow \Phi B \xrightarrow{\lambda} B \leftarrow A \to \Sigma\Omega A\ .$$

7.1.3. Projectives. Let V_n be a free $\mathcal{A}$-module on one generator having degree n. Let B_n be the submodule of V_n generated by $B(n)$. Then the cokernel in the short exact sequence

$$0 \to B_n \to V_n \to F_n \to 0$$

is an unstable $\mathcal{A}$-module; in fact,

$$F_n = \Sigma^n \mathcal{A}/B(n)\ .$$

We have

$$\mathrm{Hom}_{\mathcal{U}}(F_n, M) \cong M^n$$

where the isomorphism is via the correspondence

$$f(b_n) = x,\ b_n = s^n(1)\ .$$

We call F_n the <u>free</u> unstable $\mathcal{A}$-module on one generator of degree n. In [**74**] a proof is given that projective modules of finite type are direct sums of various F_n.

There is a canonical isomorphism

$$\Omega F_{n+1} \cong F_n$$

induced by $\sigma^* b_{n+1} = b_n$.

If we replace B_n by the submodule generated by $B(n)$ and the Bockstein,

$$\tilde{B}_n = B(n) + \mathcal{A}(\beta) \ ,$$

the resulting cokernel

$$0 \to \tilde{B}_n \to V_n \to F'_n \to 0$$

is no longer free, but we have a short exact sequence,

$$\begin{gathered} 0 \to F_{n+1} \to F_n \to F'_n \to 0 \\ b_{n+1} \mapsto \beta b_n \ . \end{gathered}$$

Moreover, we can identify $\Omega F'_{n+1}$ with F'_n as before. We write u_n, in F'_n for the image of b_n. Then

$$u_n = \sigma^* u_{n+1} \ .$$

Observe also that $\Omega_1 F_n = 0$ and $\Omega_1 F'_n = 0$ if $n \geq 2$.

7.1.4. The functor U. The forgetful functor from $\mathcal{K}$ to $\mathcal{U}$ has a left adjoint U,

$$\mathrm{Hom}_{\mathcal{U}}(M, H) \cong \ \mathrm{Hom}_{\mathcal{K}}(U(M), H)$$

for M in $\mathcal{U}$ and H in $\mathcal{K}$. Given a short exact sequence in $\mathcal{U}$,

$$0 \to A \to B \to C \to 0$$

we have

$$U(B) \cong U(A) \otimes U(C)$$

as left $U(A)$-modules, but the algebra extension may be non-trivial. In the case of direct sums, we have

$$U(M \oplus N) \cong U(M) \otimes U(N)$$

as algebras.

Another useful property is the characterization that $U(M)$ is a free commutative algebra if and only if $\Omega_1 M^+ = 0$, where $M^+ = M$ for $p = 2$ and M^+ is the even degree part for p odd.

We use U to reformulate the results of Cartan and Serre on the cohomology of Eilenberg-Mac Lane spaces [**109**].

Thus

$$\begin{aligned} H^*(K(Z/p, n); Z/p) &\cong U(F_n), \\ H^*(K(Z, n); Z/p) &\cong U(F'_n), \end{aligned}$$

and for $f \geq 2$,

$$H^*(K(Z/p^f, n); Z/p) \cong U(F'_n \oplus F'_{n+1})$$

with $\beta_f(u_n) = [\![u_{n+1}]\!]$, with zero indeterminancy.

These results include RP^∞ and CP^∞. For HP^∞, the polynomial algebra

$$Z/p[x_4]$$

is of the form $U(M)$ for $p = 2$ and is not for odd primes. In case $p = 2$, we have

$$M = \text{ span } \{x_4, Sq^4 x_4,\ Sq^8 Sq^4 x_4, \dots\} .$$

For p odd, the Adem relation $\mathcal{P}^1\mathcal{P}^1 = 2\mathcal{P}^2$ and the equation $\mathcal{P}^2 x_4 = x_4^p$ imply that

$$\mathcal{P}^1 x_4 = \lambda x_4^{\frac{p+1}{2}} \text{ with } \lambda \equiv \pm 2 \bmod p .$$

Given a projective in $\mathcal{U}$ of finite type, there is a generalized Eilenberg-Mac Lane space, written $K(P)$, and characterized up to homotopy by

$$\pi_n K(P) = \text{ Hom}_{\mathcal{U}}(P, \Sigma^n Z/p) .$$

Then $H^*(K(P); Z/p) = U(P)$, and for maps

$$f : P_1 \to P_0$$

we have $K(f) : K(P_0) \to K(P_1)$ with

$$K(f)^* = U(f) .$$

In this situation, the algebraic and geometric loop functors are closely related. The map

$$\sigma^* : P \to \Omega P$$

given by $\sigma^*(x) = s^-[x]$ induces a homotopy equivalence

$$\Omega K(P) \simeq K(\Omega P) .$$

Under this equivalence, we may identify $\Omega K(f)$ with $K(\Omega f)$.

7.1.5. The cohomology of $U(Q)$. In this section, we describe the calculation of

$$\text{Tor}^{*,*}_{U(Q)}(Z/p, Z/p)$$

in the case where $\Omega_1 Q = 0$. For $p = 2$, the result is

$$\text{Tor}^{*,*}_{U(Q)}(Z/2, Z/2) \cong E(s^{-1,0}\Sigma\Omega Q)$$

where $s^{-1,0}$ is the bi-indexed shift operator. The indexing lies in the second quadrant, with non-positive homological degrees. To describe the result for odd primes, we work with the functor Φ.

Set

$$\Phi^+ Q = \text{ submodule concentrated in degrees } \equiv 0 \bmod 2p ,$$
$$\Phi^- Q = \Phi Q/\Phi^+ Q .$$

Write $\lambda^+ : \Phi^+ Q \to Q$ for the restriction of λ. We take the quotient of λ^+ to define IQ,

$$\Phi^+ Q \xrightarrow{\lambda^+} Q \longrightarrow IQ \longrightarrow 0 .$$

The condition $\Omega_1 Q = 0$ yields a short exact sequence,

$$0 \longrightarrow \Phi^- Q \xrightarrow{\lambda^-} IQ \to \Sigma\Omega Q \to 0$$

with λ^- induced by λ and the switchback construction.

We write

$$IQ^+ = \text{submodule concentrated in even degrees,}$$

$$IQ^- = \text{submodule concentrated in odd degrees.}$$

Then we have

$$U(Q) \cong Z/p[IQ^+] \otimes E(IQ^-) .$$

The cohomology of $U(Q)$ is given by

$$\mathrm{Tor}^{*,*}_{U(Q)}(Z/p, Z/p) \cong E(s^{-1,0} IQ^+) \otimes \Gamma(s^{-1,0} IQ^-)$$

where E is an exterior algebra and Γ denotes the divided polynomial algebra. This result is proved by means of Koszul resolutions and [**105**] is a convenient reference.

7.2. Massey-Peterson fibrations

Massey and Peterson identified conditions which lead to a detailed description of the cohomology of the total space in a fibration. Their work appears in [**73**], [**74**] and is carried out for the prime 2. The odd primary case was worked out by Barcus [**8**] and Smith [**104**]. In particular, Smith employed the Eilenberg-Moore spectral sequence. In light of Smith's work, Miller and the author reworked the Massey-Peterson conditions [**42**].

Definition 7.2.1. A Massey-Peterson fibration is a fiber square

$$\begin{array}{ccc} E & \xrightarrow{h} & E_0 \\ {\scriptstyle p}\downarrow & & \downarrow{\scriptstyle p_0} \\ B & \xrightarrow[f]{} & B_0 \end{array}$$

satisfying (a)–(e);

(a): B is connected and B_0 is either simply connected, or B and B_0 are connected H-spaces and f is an H-map, p_0 is a contractible fibration.

(b): The mod p cohomologies H^*B, H^*B_0 are of finite type over Z/p.

(c): H^*B is a free module over the subalgebra $im f^*$

(d): There is an $\mathcal{A}$-submodule Q in $\bar{H}^*B_0$ such that $\Omega_1 Q = 0$ and H^*B_0 is free over $U(Q)$.

(e): $\ker f^*$ is the ideal generated by Q.

The principal (but by no means only) example is the stable two-stage Postnikov system described below. Let

$$f : P_1 \to P_0$$

be a map of finitely generated projectives in $\mathcal{U}$. We form the fiber square

$$\begin{array}{ccc} W & \longrightarrow & PK(P_1) \\ \downarrow & & \downarrow \\ K(P_0) & \xrightarrow[K(f)]{} & K(P_1) \end{array},$$

the homotopy theoretic fiber of $K(f)$. Then both (a) and (b) hold. Indeed, $K(f)$ is an H-map with respect to the canonical (product) H-structures on $K(P_0)$, $K(P_1)$. To check (c), we look at $\ker f$ and $\operatorname{coker} f$,

$$0 \leftarrow M \leftarrow P_0 \overset{f}{\leftarrow} P_1 \leftarrow Q \leftarrow 0 .$$

Then we have a short exact sequence in $\mathcal{U}$

$$0 \leftarrow M \leftarrow P_0 \leftarrow im\ f \leftarrow 0 .$$

and $U(P_0) \cong U(M) \otimes U(im\ f)$.

Thus (c) is satisfied. Since $\Omega_1 P_1 = 0$, the same holds for Q. Moreover, from the short exact sequence

$$0 \leftarrow im\ f \leftarrow P_1 \leftarrow Q \leftarrow 0 ,$$

we obtain $U(P_1) \cong U(im\ f) \otimes U(Q)$.

Thus both (d) and (e) hold.

Another example is provided by $\mathrm{Spin}(n)$, the universal covering space of $SO(n)$. We have the fiber square

$$\begin{array}{ccc} \mathrm{Spin}(n) & \longrightarrow & PK \\ \downarrow & & \downarrow \\ SO(n) & \xrightarrow{f} & K(Z/2, 1) \end{array},$$

with

$$\begin{aligned} H^*(SO(n);\ Z/2) &\cong U(\bar{H}^* RP^{n-1}) \\ &\cong Z/2[x_1]/(x_1^{2^k} = 0) \otimes C \end{aligned}$$

where $2^{k-1} < n \leq 2^k$ determines k and f maps H^1 isomorphically. In this case

$$Q = \text{ Span } \{Sq(2^{k-1}, \ldots, 1)b_1, Sq(2^k, \ldots, 1)b_1, \ldots\} .$$

7.2.1. Structure Theorems. We establish two structure theorems for the cohomology of Massey-Peterson fibrations. We first introduce the algebra

$$R = H^*B \otimes_{H^*B_0} Z/p$$

where the action is induced by f^*. In our two examples,

$$R = U(M),\ R = C .$$

The first structure theorem asserts that under the hypotheses (a)–(e)

$$(i) \qquad H^*(E) \cong R \otimes U(\Omega Q) \text{ with } \otimes \text{ over } Z/p$$

and

$$(ii) \qquad R \cong H^*(B) / \ker p^* .$$

The isomorphism in part (i) is as R-modules. In particular, this isomorphism need not lie in the category $\mathcal{K}$ or even $\mathcal{U}$. The issue is the extension of the Steenrod action from ΩQ over the tensor product. Massey and Peterson introduce a certain short exact sequence which embraces the extension issue. Meanwhile, if $\Omega_1 \Omega Q = 0$, then (i) holds as algebras, since in this case $U(\Omega Q)$ is a free algebra.

The first structure theorem employs the Thom complex. For the fiber square,

$$\xi : \quad \begin{array}{ccc} E & \xrightarrow{h} & E_0 \\ {\scriptstyle p}\downarrow & & \downarrow{\scriptstyle p_0} \\ B & \xrightarrow[f]{} & B_0 \end{array} ,$$

let E_T and E_{0T} denote the unreduced mapping cylinders of p and p_0 respectively. Let

$$T(\xi) : (E_T, E) \longrightarrow (E_{0T}, E_0)$$

be the map induced by h. Under the assumption that E_0 is contractible, we have

$$H^*(E_{0T}, E_0) \cong \bar{H}^*B_0$$

and the following diagram in $\mathcal{U}$,

$$\begin{array}{ccccccccc} & & 0 & \longrightarrow & \bar{H}^*B_0 & \longrightarrow & H^*B_0 & \longrightarrow & Z/p & \longrightarrow & 0 \\ & & & & \downarrow{\scriptstyle T(\xi)^*} & & \downarrow{\scriptstyle f^*} & & \downarrow & & \\ & & H^{*-1}(E) & \xrightarrow[\delta]{} & H^*(E_T, E) & \longrightarrow & H^*B & \xrightarrow[p^*]{} & H^*E . & & \end{array}$$

For dimensional reasons, $im\ \delta$ is a suspension in $\mathcal{U}$, so there is a well-defined unstable $\mathcal{A}$-module $M(\xi)$ given by

$$\Sigma M(\xi) = im\ \delta \cap im\ T(\xi)^* \ .$$

We define $N(\xi)$ as the pull-back

$$N(\xi) = \delta^{-1}\Sigma M(\xi) \ .$$

Thus we have a short exact sequence in $\mathcal{U}$,

$$0 \to \ker \delta \to N(\xi) \to M(\xi) \to 0 \ .$$

In the proof of the first structure theorem, we shall establish

$$\textit{(iii)} \qquad M(\xi) \cong R \otimes \Omega Q, \ \textit{as } R\textit{-modules.}$$

Before we give the second structure theorem in detail, we discuss its context. Essentially, the second structure theorem extends the scope of the Serre exact sequence beyond the "stable range" restriction. The Serre exact sequence embraces the extension issue for the Steenrod action in terms of the transgression operator,

$$H^{*-1}F \xrightarrow{\tau} H^*B, \ F = \text{ fiber } \simeq \Omega B_0 \ ,$$

and a short exact sequence

$$0 \to \text{ coker } \tau \to H^*E \to \ker \tau \to 0$$

in a range of dimensions. *The second structure theorem for Massey-Peterson fibrations is the following diagram in $\mathcal{U}$, the top rows are short exact sequences,*

$$\begin{array}{ccccccccc}
0 & \longrightarrow & R & \longrightarrow & G(\xi) & \longrightarrow & \Omega Q & \longrightarrow & 0 \\
 & & \| & & \downarrow & & \downarrow & & \\
0 & \longrightarrow & R & \longrightarrow & N(\xi) & \longrightarrow & M(\xi) & \longrightarrow & 0 \\
 & & & & \downarrow & & \downarrow & & \\
 & & & & H^*(E) & \underset{i^*}{\longrightarrow} & H^*(F) \ . & &
\end{array}$$

The map $\Omega Q \to M(\xi)$ is given by $x \to 1 \otimes x$ under the isomorphism of (iii). The module $G(\xi)$ is the pull-back. The top row is called the <u>*fundamental sequence*</u> *of the Massey-Peterson fibration. In light of the first structure theorem, the extension issue in the fundamental sequence includes the extension issue for H^*E as an algebra over the Steenrod algebra.* We turn to proofs of the two structure theorems.

Step 1. We calculate E_2 for the Eilenberg-Moore spectral sequence

$$E_2 = \ \text{Tor}^{*,*}_{H^*(B_0)}(H^*B, Z/p) \ .$$

Write $A = im\ f^*$. We have the Cartan-Eilenberg spectral sequence ([**23**], p. 348)

$$\mathrm{Tor}^*_A(H^*B, \mathrm{Tor}_{H^*B_0}(A, Z/p)) \Rightarrow E_2 \ .$$

Since H^*B is free over A, the Cartan-Eilenberg spectral sequence collapses and we have

$$H^*B \otimes_A \ \mathrm{Tor}^*_{H^*(B_0)}(A, Z/p) \cong E_2 \ .$$

Furthermore, $H^*B \cong (H^*B//A) \otimes A$ and

$$H^*B_0 \cong A \otimes U(Q) \ .$$

Hence the functors $H^*B \otimes_A -$ and $H^*B//A \otimes -$ give the same answers and we may write

$$H^*B//A \cong H^*B \otimes_A Z/p \cong H^*B \otimes_{H^*B_0} Z/p = R \ .$$

Hence the E_2-term is

$$R \otimes \ \mathrm{Tor}^*_{H^*(B_0)}(A, Z/p)$$

with R concentrated in homological degree 0.

To calculate the remaining Tor term, we have

$$H^*B_0 \cong A \otimes U(Q) \ .$$

Hence

$$\mathrm{Tor}^*_{H^*B_0}(A, Z/p) \cong \mathrm{Tor}^*_{U(Q)}(Z/p,\ Z/p) \ .$$

We recall the calculation from subsection (7.1.5) to write

$$\mathrm{Tor}^*_{U(Q)}(Z/p, Z/p) \cong E(s^{-1,0}IQ^+) \otimes \Gamma(s^{-1,0}IQ^-) \ .$$

Step 2. We use Smith's calculation for differentials [**105**]. The result is (for odd primes p)

$$E^*_p \cong E^*_\infty \cong R \otimes E(s^{-1,0}(\Sigma\Omega Q)^+) \otimes T_p[s^{1,0}(\Sigma\Omega Q)^-]$$

where T_p indicates the polynomial algebra truncated at height p.

Step 3. Extension from E^*_∞ to H^*E.

We observe

$$s^{-1,0}(\Sigma\Omega Q)^+ \cong (\Omega Q)^- \ ,$$
$$s^{-1,0}(\Sigma\Omega Q)^- \cong (\Omega Q)^+ \ .$$

Hence we have a map $\Omega Q \to E^{-1,*}_\infty$ and an extension

$$U(\Omega Q) \to E \otimes T_p \ .$$

This extension is an isomorphism as seen by filtering $U(\Omega Q)$ by powers,

$$F^0 = Z/p,\ F^{-1} = Z/p \otimes \Omega Q \ ,$$
$$F^{-n} = (F^{-1})^n \ .$$

Then $F^{-p} = U(\Omega Q)$. Thus $H^*E \cong R \otimes U(\Omega Q)$ as R-modules, completing the proof of for part (i) of the first structure theorem, for odd primes. For $p = 2$, $E_2^{*,*} = E_\infty^{*,*}$ and the extension step works as before.

For part (ii), we observe that R maps surjectively to $H^*B/\ker p^*$ by its construction, since $f \circ p$ is null-homotopic. The calculation of $E_\infty^{*,*}$ yields that the natural projection is an isomorphism.

To work out the structure of the fundamental sequence, we first recall the diagram used for the definition of $M(\xi)$,

$$\begin{array}{ccccccc}
0 & \longrightarrow & \bar{H}^*B_0 & \longrightarrow & H^*B_0 & \longrightarrow & Z/p & \longrightarrow & 0 \\
 & & \Big\downarrow{\scriptstyle T(\xi)^*} & & \Big\downarrow{\scriptstyle f^*} & & \Big\downarrow & & \\
H^{*-1}(E) & \xrightarrow{\delta} & H^*(E_T, E) & \longrightarrow & H^*B & \xrightarrow[p^*]{} & H^*E \,. & &
\end{array}$$

Application of the functor $H^*B \otimes_{H^*B_0} -$ to the top row produces the following diagram of exact sequences,

$$\begin{array}{ccccccccc}
 & & & & & & & & 0 \\
 & & & & & & & & \Big\downarrow \\
0 & \longrightarrow & \mathrm{Tor}^{-1}_{H^*B_0}(H^*B, Z/p) & \longrightarrow & H^*B \otimes_{H^*B_0} \bar{H}^*B_0 & \longrightarrow & H^*B & \longrightarrow & R \\
 & & \Big\downarrow{\scriptstyle (a)} & & \Big\downarrow{\scriptstyle (b)} & & \Big\| & & \Big\downarrow{\scriptstyle p^*} \\
0 & \longrightarrow & M(\xi) & \longrightarrow & M'(\xi) & \longrightarrow & H^*B & \xrightarrow[p^*]{} & H^*E \,.
\end{array}$$

Here we have written R for $H^*B \otimes_{H^*B_0} Z/p$ and used part (ii) to write the displayed monomorphism on the right. Thus, diagram chasing yields the equality

$$\ker p^* = im\ H^*B \otimes_{H^*B_0} T(\xi)^* \ ,$$

and we have written $M'(\xi)$ for the latter module. Similarly, the two vertical maps on the left are surjections.

We now look at the right surjection (b). The $\mathcal{A}$-submodule

$$1 \otimes \Phi^+ Q$$

maps to 0 in $H^*B \otimes_{H^*B_0} \bar{H}^*B_0$, because the elements are p-th powers of elements in $U(Q)$. The calculation of $E_2^{*,*}$ yields the information that the mapping

$$R \otimes IQ \to M'(\xi)$$

is a momomorphism. Smith's calculation of differentials includes the following diagram:

$$\begin{array}{ccccccc}
E_{p-1}^{-p,*} & \xrightarrow{d_{p-1}} & E_{p-1}^{-1,*} & \longrightarrow & E_p^{-1,*} & = & E_\infty^{-1,*} \\
\| & & \downarrow = & & & & \downarrow \cong \\
R\otimes \gamma_p(s^{-1,0}IQ^-) & & & & & & \\
\| & & & & & & \\
R\otimes s^{-2,0}\Phi^- Q & \xrightarrow[1\otimes s\lambda^-]{} & R\otimes s^{-1,0}IQ & \longrightarrow & & & R\otimes s^{-1,0}\Sigma\Omega Q \longrightarrow 0
\end{array}$$

Hence, the left vertical surjection (a) induces a well-defined map in $\mathcal{U}$

$$R\otimes s^{-1,0}\Sigma\Omega Q \to M(\xi),$$

and this map is surjective.

We can now establish (iii). We have factored the connecting homomorphism δ in the diagram below:

$$\begin{array}{ccccccc}
E_\infty^{-1,\infty} & \xleftarrow{\cong} & R\otimes s^{-1,0}\Sigma\Omega Q & \twoheadrightarrow & M(\xi) & \rightarrowtail & M'(\xi) \\
\uparrow & & & & & & \downarrow \\
F^{-1,*} & & & & & & \\
\downarrow & & & & & & \\
H^{*-1}(E) & & & \xrightarrow{\delta} & & & H^*(E_T, E)
\end{array}$$

We have $\ker\delta \cong R \cong E_\infty^{0,*}$. Hence the mapping

$$\delta : F^{-1,*} \to im\ \delta$$

is an isomorphism. Hence the map

$$R\otimes s^{-1,0}\Sigma\Omega Q \to M(\xi)$$

is monic, completing the proof of (iii).

The fundamental sequence is obtained using (iii) to make the pull-back construction, and was described earlier. We display the edge homomorphisms, as the final detail of the discussion,

$$\begin{array}{ccccccc}
H^*B & \longrightarrow & H^*B//im\ f^* = R(\xi) & \longrightarrow & G(\xi) & \longrightarrow & \Omega Q(\xi) = s^{-1,0}\Sigma\Omega Q \\
\| & & & & \downarrow & & \downarrow \\
H^*(B) & & \xrightarrow[p^*]{} & & H^*(E) & \xrightarrow[i^*]{} & H^*(F)\ .
\end{array}$$

The identification of i^* follows by naturality of the Eilenberg-Moore spectral sequence applied to the diagram

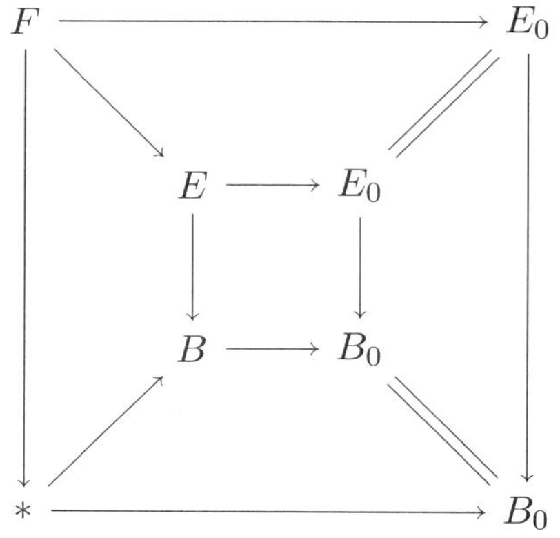

and

$$\begin{array}{ccc} H^{*-1}(F) & \xrightarrow{\cong} & H^*(CF, F) \\ \uparrow{\scriptstyle i^*} & & \uparrow \\ H^{*-1}(E) & \longrightarrow & H^*(E_T, E) \end{array}$$

□

Our first use of the fundamental sequence is to answer a question raised in Chapter 4, concerning the representation of secondary operations by coliftings.

We observe

(*iv*) $\quad$ *$G(\xi)$ is represented by coliftings.*

To see this, let e be an element of $G(\xi)$ satisfying

$$i^*(e) = \sigma^* x$$

for x in $Q \subset \ker f^*$. We represent x by a map

$$g : B_0 \to K(Z/p, |e| + 1) .$$

Then we have a colifting γ displayed in the diagram below:

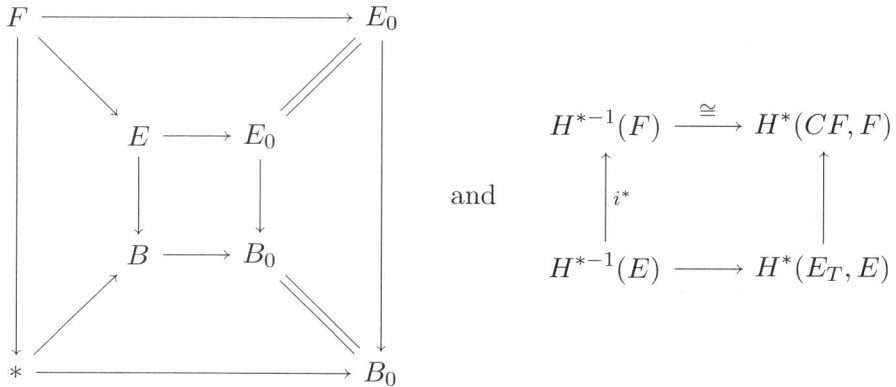

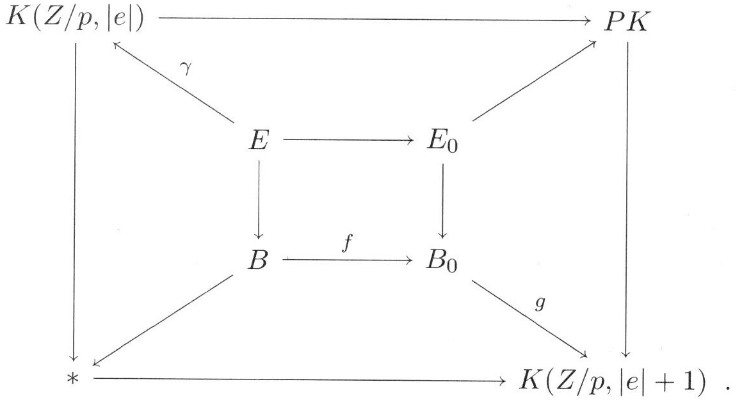

Since $i^*|e| = \sigma^* x = i^* \gamma^* b_{|e|}$, and by construction

$$\gamma^* b_{|e|} \text{ is in } G(\xi) ,$$

there is $\delta : B \to K(Z/p, |e|)$ such that

$$e - \gamma^* b_{|e|} = p^* \delta \, .$$

This element may be used to alter the null-homotopy for $g \circ f$ so as to obtain the element e. $\square$

The secondary operation represented by e has the form

$$\Theta : S_f(\quad) \to T_{\Omega g}(\quad) \, ,$$

using compatibility of coliftings with the principal action (3.2.8). For $\epsilon : X \to B$ with $f \circ \epsilon \sim *$, we have

$$\Theta(\epsilon) = [\![\bar{\epsilon}^*(e)]\!] \, .$$

In the literature, Θ is often recognized up to tethering by its restriction to the fiber in its universal example.

7.2.2. Coproduct theorem. Here we develop coalgebra information for the case of the cohomology of stable two-stage Postnikov systems. We are reworking results published in [**27**], [**38**], [**43**], [**54**], and [**78**]. Suppose we have the system

$$\begin{aligned} B &= K(P_0), \; B_0 = K(P_1) \\ \theta &: B \to B_0 \text{ given by } \theta = K(f) \\ \text{for } f &: P_1 \to P_0 \, . \end{aligned}$$

Then we have the pair of Massey-Peterson fibrations

$$\xi : \quad \begin{array}{ccc} W_\theta & \longrightarrow & PB_0 \\ {\scriptstyle p}\big\downarrow & & \big\downarrow \\ B & \xrightarrow[\theta]{} & B_0 \end{array} \, , \qquad \Omega\xi : \quad \begin{array}{ccc} \Omega W_\theta & \longrightarrow & \Omega PB_0 \\ {\scriptstyle \Omega p}\big\downarrow & & \big\downarrow \\ \Omega B & \xrightarrow[\Omega\theta]{} & \Omega B_0 \end{array} \, .$$

The fundamental sequence for $\Omega\xi$ is

$$0 \to U(\operatorname{coker} \Omega f) \to G(\Omega\xi) \to \Omega(\ker \Omega f) \to 0 \, .$$

Moreover, except for the case $P_0 = F_2'$, we have

$$\Omega_1(im\, \Omega f) = 0 \, .$$

A consequence is that

$$\Omega(\ker \Omega f) \to \Omega^2 P_1$$

is monic. This follows from the exact sequence obtained from

$$0 \to \ker \Omega f \to \Omega P_1 \to im\, \Omega f \to 0$$

by application of the algebraic loop functor. In particular. we may write elements of $\Omega(\ker \Omega f)$ unambiguously in terms of a basis for $\Omega^2 P_1$, and receive them from ΩP_1 under the algebraic suspension,

$$\begin{array}{ccc} \Omega(\ker \Omega f) & \longrightarrow & \Omega^2 P_1 \\ \uparrow & & \uparrow \\ \ker \Omega f & \longrightarrow & \Omega P_1 \ . \end{array}$$

The information needed to state the coproduct theorem is developed in the following diagram:

$$\begin{array}{ccccccc} & & & & & & \Phi P_0 \\ & & & & & & \downarrow \lambda \\ & & & & P_1 & \xrightarrow{f} & P_0 \\ & & & & \downarrow \sigma^* & & \downarrow \sigma^* \\ \Omega(\ker \Omega f) & \longleftarrow & \ker \Omega f & \hookrightarrow & \Omega P_1 & \xrightarrow{\Omega f} & \Omega P_0 \ . \end{array}$$

Given x in $\Omega(\ker \Omega f)$, we may lift it to x_1 in $\ker \Omega f$ and x_2 in ΩP_1. Then we may write $x_2 = \sigma^* x_3$ and $f(x_3) = \lambda \bar{x}_4$. In particular, if $p = 2$ and $|x|$ is odd, then $f(x_3) = 0$. If $|x|$ is even, then

$$f(x_3) = Sq^{n+1} x_4 \text{ with } 2n = |x|,\ (n+1) = |x_4| \ .$$

If p is odd and $|x| = 2np$, we have

$$f(x_3) = \beta \mathcal{P}^n x_4 \text{ with } (2n+1) = |x_4| \ .$$

For other degrees, we get 0. The choices involved in picking x_1, x_3 do not affect the value of $\sigma^* x_4$. We illustrate this procedure with an example,

$$\Omega B = K(Z, 3) \xrightarrow{\mathcal{P}^p \mathcal{P}^1} K(Z/p, 2p^2 + 1) = \Omega B_0 \ .$$

The Adem relation

$$\beta \mathcal{P}^1 \beta (\mathcal{P}^p \mathcal{P}^1) = \beta \mathcal{P}^{p+1} \beta \mathcal{P}^1$$

yields the fact that this composite is 0 on 3-dimensional classes but non-zero on 4-dimensional classes. In this case

$$x = \beta \mathcal{P}^1 \beta b_{2p^2} \text{ in } H^*(K(Z/p, 2p^2)) \ .$$

We may take $x_1 = \beta \mathcal{P}^1 \beta b_{2p^2+1} = x_2$, and $x_3 = \beta \mathcal{P}^1 \beta b_{2p^2+2}$. Then $x_4 = \beta \mathcal{P}^1 b_4$ and

$$\sigma^* x_4 = -\beta \mathcal{P}^1 b_3 \ .$$

The statement of the coproduct theorem is the following. *If $p = 2$ and $|x|$ is even, then there exists*

$$e \text{ in } G(\Omega\xi), \text{ projecting to } x$$

with coproduct

$$\text{(i)} \qquad 1 \otimes e + e \otimes 1 + \Omega p^*(\sigma^* x_4 \otimes \sigma^* x_4)\ .$$

For odd primes, if $|x| = 2np$, there is e projecting to x with coproduct

$$\text{(ii)} \qquad 1 \otimes e + e \otimes 1 + \Omega p^* \left(\sum_{j=1}^{p-1} \frac{1}{p} \binom{p}{j} (\sigma^* x_4)^j \otimes (\sigma^* x_4)^{p-j} \right)\ .$$

In other dimensions, e may be chosen to be a primitive element.

The proof amounts to translating the information developed in Chapter 3 into the present notation. We start by representing x_2 in $\Omega P_1 \cap \ker \Omega f$ by a map

$$\Omega\varphi : \Omega K(P_1) \to \Omega K(Z/p,\ |x| + 2)\ .$$

Then the image of x in $\Omega^2 P_1$ is represented by $\Omega^2\varphi$. In the diagram below, $\varphi\theta$ represents $\lambda\bar{x}_4 = \theta x_3$,

$$\begin{array}{ccccc}
B_2 W_\theta & \longrightarrow & (\Sigma\Omega W_\theta)^{(2)} & \longrightarrow & \Sigma^2\Omega W_\theta \\
\downarrow & & \downarrow & & \\
B_2 & \longrightarrow & (\Sigma\Omega K(P_0))^{(2)} & & \Big\downarrow \gamma \\
\downarrow & & \Big\downarrow \zeta & & \\
K(P_0) & \xrightarrow[\varphi\theta]{} & K(Z/p, |x| + 2) & = & K(Z/p, |x| + 2)\ .
\end{array}$$

Now $\varphi\theta$ represents $\beta P^n x_4$ with $|x_4| = 2n + 1$, for the case of odd primes. In this case, the signed Bockstein is (-1) times the ordinary Bockstein and the minus sign in (3.9.5) is cancelled. Hence we have

$$\zeta = \sum_{j=1}^{p-1} \frac{1}{p} \binom{p}{j} \left[(\sigma^* x_4)^j \mid (\sigma^* x_4)^{p-j} \right]\ .$$

□

We give another example where signs are involved. Consider the situation

$$\Omega B = K(Z/p, 4) \xrightarrow{\mathcal{P}^1} K(Z/p, 2p + 2) = \Omega B_0\ .$$

Now $\beta\mathcal{P}_4^2 = 0$ but $\beta\mathcal{P}^2 b_5 \neq 0$. We have the Adem relation

$$\beta\mathcal{P}^1(\mathcal{P}^1) = 2\beta\mathcal{P}^2 .$$

So we have $x = \sigma^*(x_2)$ with $x_2 = \beta\mathcal{P}^1 b_{2p+2}$. Then we obtain

$$\begin{aligned} x_3 &= -\beta\mathcal{P}^1 b_{2p+3} , \\ x_3 &\to -\beta\mathcal{P}^1(\mathcal{P}^1 b_5) = \beta\mathcal{P}^1(-2b_5) , \\ \text{so } x_4 &= -2b_5 , \\ \text{and } \sigma^* x_4 &= -2b_3 . \end{aligned}$$

In terms of maps and the unsigned Bockstein, we have the diagram

$$\begin{array}{ccccc} K(Z/p, 2p+1) & & & & \\ \downarrow & \searrow^{-\beta\mathcal{P}^1} & & & \\ \Omega W & \xrightarrow{\widetilde{B\mathcal{P}^1}} & K(Z/p, 4p) & & \\ \downarrow & & & & \\ K(Z/p, 4) & \xrightarrow[\mathcal{P}^1]{} & K(Z/p, 2p+2) & \xrightarrow[\beta\mathcal{P}^1]{} & K(Z/p, 4p+1) . \end{array}$$

Remark. Let $M =$ coker f for $f : P_1 \to P_0$. The construction used for the coproduct theorem yields a map

$$\Omega\Omega_1 M \to \Omega M$$

sending elements in degree $2np$ $(2n)$ to elements in degree $2n$ (n) for p odd and $p = 2$ respectively. In terms of that construction, the map is given by

$$[x] \to [\sigma^* x_4] ,$$

where $\Omega\Omega_1 M$ is a quotient of $\Omega(\ker \Omega f)$ and ΩM is a quotient of ΩP_0.

The case $\sigma^* x_4 = 0$ is not excluded from our statement of the coproduct theorem. Two examples where this happens are, for $p = 2$, W is the homotopy fiber of

$$K(Z_2, 2) \xrightarrow{Sq^4 Sq^2} K(Z/2, 8) .$$

Then $\Omega W \simeq K(Z/2, 1) \times K(Z/2, 6)$ with b_6 primitive. This equivalence is as H-spaces, but not (as it turns out) as loop spaces.

For p odd, setting W to be the homotopy fiber of

$$K(Z/p, 2n) \xrightarrow{\mathcal{P}^n} K(Z/p, 2np)$$

results in $\Omega W \simeq K(Z/p, 2n-1) \times K(Z/p, 2np-2)$ as H-spaces.

We make an application of the coproduct theorem to complete the discussion begun in Chapter 6 concerning the existence of a space with mod 2 cohomology

$$Z/2[w]/w^4 \text{ with } |w| = 8 .$$

The argument in Chapter 6 ruled out such a space by factoring w^3 through a secondary operation and comparing this factorization with Adams' decomposition of Sq^{16}. The details concerning the factorization are given here.

Let $\Phi'_{0,3}$ be a secondary operation based on $[\varphi][\theta] = 0$ where

$$\varphi = (Sq(8) + Sq(6,2),\ Sq(7) + Sq(4,2,1),\ Sq(1)),$$
$$\theta = col(Sq(1), Sq(2), Sq(8)) .$$

Let Ψ be a secondary operation based on the relation

$$Sq^1 Sq^{12} = Sq^{13} .$$

In particular, Ψ is defined on 12-dimensional classes with cup square equal to 0.

Let x be an 8-dimensional mod 2 reduction of an integral class with $Sq^{12}Sq^4 x = 0$. Then $\Phi'_{0,3}$ is defined on $Sq^8 x$ and we have

(iii) $$\Phi'_{0,3} Sq^8 x = [\![x^3 + Sq^2 x \cup Sq^4 Sq^2 x + x \cup Sq^6 Sq^2 x]\!] + \Psi Sq^4 x + \delta$$

where δ is in span $\{Sq^{11}Sq^5 x, Sq^{10}Sq^4Sq^2 x\}$ and the value is taken modulo

$$\text{ind } (\Phi'_{0,3}, X) \supset \text{ ind } (\Psi, X) .$$

Proof. We set up the following diagram, with spaces and maps understood as loop spaces and loop maps,

$$\begin{array}{ccccccc} E & \xrightarrow{h} & W_\theta & \xrightarrow{\tilde{\varphi}} & K(24) & & \\ \downarrow & & \downarrow p & & & & \\ K(Z,8) & \xrightarrow[Sq(8)]{} & K(16) & \xrightarrow[\theta]{} & K(17,18,24) & \xrightarrow[\varphi]{} & K(25) . \end{array}$$

Here, E is the pull-back of the fibration p by $Sq(8)$.

Calculating, we have

$$\theta Sq(8) u_8 = \begin{bmatrix} Sq^9 u_8 \\ Sq^{10} u_8 \\ (Sq^{14}Sq^2 + Sq^{12}Sq^4) u_8 \end{bmatrix} = \begin{bmatrix} 0 \\ 0 \\ Sq^{12}Sq^4 u_8 \end{bmatrix} .$$

Hence E has the homotopy type of

$$E' \times K(16) \times K(17)$$

where E' is the homotopy fiber of $Sq^{12}Sq^4$. We write f_{16}, f_{17} respectively for the fundamental classes of $K(16), K(17)$ resp. We use the coproduct theorem to describe the loop structure of E in terms of the displayed splitting.

We can choose f_{16} to have coproduct

$$1 \otimes f_{16} + f_{16} \otimes 1 + u_8 \otimes u_8$$

and f_{17} to be primitive, by using the $Sq^9 u_8$, $Sq^{10} u_8$ information.

The composition $\tilde{\varphi} \circ h$ is a loop map representing a primitive class and in terms of the splitting of E is given by

$$(Sq^8 + Sq^6 Sq^2) f_{16} + (Sq^7 + Sq^4 Sq^2 Sq^1) f_{17} + e + \text{ terms from } H^* K(Z, 8)$$

with e restricting to $Sq^4 f_{23}$ in the fiber. Using the coproduct theorem, we may choose e to have coproduct

$$1 \otimes e + e \otimes 1 + Sq^4 u_8 \otimes Sq^4 u_8$$

because e arises from the relation

$$Sq^1 (Sq^{12} Sq^4) = Sq^{13} Sq^4 \ .$$

Expanding $(Sq^8 + Sq^6 Sq^2) f_{16} + e$ under the coproduct forces the cup product terms in (iii), and determines these summands up to primitives from the cohomology of $K(Z, 8)$. In dimension 24, these primitives are those indicated. Finally, in choosing e as above, we may do so through the factorization

$$\begin{array}{ccc} K(Z,8) & \xrightarrow{Sq^4} & K(12) \\ {\scriptstyle Sq^{12}Sq^4}\downarrow & & \downarrow{\scriptstyle Sq^{12}} \\ K(24) & = & K(24) \end{array}$$

and applying the coproduct theorem in the homotopy fibre of Sq^{12}. □

Exercise 7.2.2. Let $\varphi_{0,2}$ be based on the relation

$$Sq^4 Sq^1 + (Sq^2 Sq^1) Sq^2 + Sq^1 Sq^4 = 0 \ .$$

Let Ψ be based on the relation

$$Sq^1 Sq^6 = Sq^7 \ .$$

Then if x is a 4-dimensional mod 2 reduction of an integral class and $Sq^6 Sq^2 x = 0$, we have

$$\varphi_{0,2} Sq^4 x = [\![x^3]\!] + \Psi Sq^2 x \ ,$$

with the value taken modulo Ind $(\varphi_{0,2}, X)$.

The formula may be applied to show that no suspension of HP^3 splits as a non-trivial bouquet.

7.2.3. Naturality properties of fundamental sequences. We develop the naturality properties for maps of Massey-Peterson fibrations and for their loop spaces.

Suppose we are given two Massey-Peterson fiber squares satisfying the conditions of (7.2.1),

$$\xi : \quad \begin{array}{ccc} E & \xrightarrow{h} & E_0 \\ {\scriptstyle p}\downarrow & & \downarrow{\scriptstyle p_0} \\ B & \xrightarrow[f]{} & B_0 \end{array} \qquad \xi' : \quad \begin{array}{ccc} E' & \xrightarrow{h'} & E'_0 \\ {\scriptstyle p'}\downarrow & & \downarrow{\scriptstyle p'_0} \\ B' & \xrightarrow[f']{} & B'_0 \end{array} .$$

Suppose also that there is a homotopy commutative diagram

$$\begin{array}{ccc} B & \xrightarrow{f} & B_0 \\ {\scriptstyle g}\downarrow & & \downarrow{\scriptstyle g_0} \\ B' & \xrightarrow[f']{} & B'_0 \end{array} .$$

The corresponding naturality property is the following commutative diagram of fundamental sequences, under the additional assumption that

$$g_0^* : Q(\xi') \to Q(\xi) ;$$

$$\text{(i)} \qquad \begin{array}{ccccccccc} 0 & \longrightarrow & R(\xi') & \longrightarrow & G(\xi') & \longrightarrow & \Omega Q(\xi') & \longrightarrow & 0 \\ & & {\scriptstyle g^*}\downarrow & & \downarrow & & \downarrow{\scriptstyle \Omega g_0^*} & & \\ 0 & \longrightarrow & R(\xi) & \longrightarrow & G(\xi) & \longrightarrow & \Omega Q(\xi) & \longrightarrow & 0 \end{array} .$$

Proof. Since both E_0 and E'_0 are contractible, there is a map

$$\bar{g}_0 : E_0 \to E'_0$$

such that $p'_0\bar{g}_0 = g_0 p_0$. Then the composition $f'gp$ is homotopic to $g_0 f p = p'_0\bar{g}_0 h$. By the homotopy lifting property, there is a map

$$\gamma : E \to E'$$

such that $p'\gamma = gp$ and $h'\gamma \simeq \bar{g}_0 h$ by a homotopy covering a given homotopy for $g_0 f \simeq f'g$. Hence, we have the following homotopy commutative diagram of pairs,

$$\begin{array}{ccc} (E_T, E) & \xrightarrow{T(\xi)} & (E_{0T}, E_0) \\ {\scriptstyle T(g)}\Big\downarrow & & \Big\downarrow{\scriptstyle T(g_0)} \\ (E'_T, E') & \xrightarrow[T(\xi')]{} & (E'_{0T}, E'_0) \end{array} ,$$

with $T(g)|E = \gamma$, $T(g_0)|E_0 = \bar{g}_0$. The constructions in subsection (7.2.1) yield the following ladder of exact sequences,

$$\begin{array}{ccccccccc} 0 & \longrightarrow & R(\xi') & \longrightarrow & N(\xi') & \longrightarrow & M(\xi') & \longrightarrow & 0 \\ & & {\scriptstyle g^*}\Big\downarrow & & {\scriptstyle \gamma^*}\Big\downarrow & & \Big\downarrow{\scriptstyle T(g)^*} & & \\ 0 & \longrightarrow & R(\xi) & \longrightarrow & N(\xi) & \longrightarrow & M(\xi) & \longrightarrow & 0 . \end{array}$$

Under the hypothesis on g_0^*, we may apply the two structure theorems to obtain (i). □

For loop spaces, we impose the assumption that both ξ and $\Omega\xi$ are Massey-Peterson fibrations,

$$\Omega\xi : \quad \begin{array}{ccc} \Omega E & \xrightarrow{\Omega h} & \Omega E_0 \\ {\scriptstyle \Omega p}\Big\downarrow & & \Big\downarrow{\scriptstyle \Omega p_0} \\ \Omega B & \xrightarrow[\Omega f]{} & \Omega B_0 \end{array}$$

and the cohomology suspension $\sigma^* : H^*(B_0) \to H^*(\Omega B_0)$ satisfies

$$\sigma^* : Q(\xi) \to Q(\Omega\xi) .$$

Under these conditions, we have the following ladder of fundamental sequences,

$$\text{(ii)} \quad \begin{array}{ccccccccc} 0 & \longrightarrow & R(\xi) & \longrightarrow & G(\xi) & \longrightarrow & \Omega Q(\xi) & \longrightarrow & 0 \\ & & {\scriptstyle \sigma^*}\Big\downarrow & & \Big\downarrow & & \Big\downarrow{\scriptstyle \sigma^*} & & \\ 0 & \longrightarrow & R(\Omega\xi) & \longrightarrow & G(\Omega\xi) & \longrightarrow & \Omega Q(\Omega\xi) & \longrightarrow & 0 \end{array}$$

and σ^* on the right is the algebraic loop map.

Proof. The fundamental sequence for $\Omega\xi$ is extracted from

$$T(\Omega\xi) : ((\Omega E)_T, \Omega E) \longrightarrow ((\Omega E_0)_T, \Omega E_0) .$$

We regard $(\Omega E)_T$ as a push-out,

$$\begin{array}{ccc} \Omega E & \longrightarrow & \Omega E \times I \\ \downarrow & & \downarrow \\ \Omega B & \longrightarrow & (\Omega E)_T \ . \end{array}$$

We have a map $\Omega E \times I \to \Omega(E \times I)$ obtained as the adjoint to evaluation on the first factor,

$$\Sigma(\Omega E \times I) \longrightarrow E \times I \ ,$$
$$(t, x, s) \longrightarrow (x(t), s) \ .$$

By naturality of push-outs, there is a homotopy equivalence

$$w : (\Omega E)_T \longrightarrow \Omega(E_T) \ .$$

Doing the same for E_0 yields a commutative diagram

$$\begin{array}{ccc} ((\Omega E)_T, \Omega E) & \xrightarrow{T(\Omega f)} & ((\Omega E_0)_T, \Omega E_0) \\ \downarrow{\scriptstyle w} & & \downarrow{\scriptstyle w_0} \\ (\Omega E_T, \Omega E) & \xrightarrow[\Omega T(f)]{} & (\Omega E_{0T}, \Omega E_0) \ . \end{array}$$

Thus we may use $\Omega T(f)$ to construct the fundamental sequence for $\Omega\xi$.

Now we have the commutative diagram

$$\begin{array}{ccc} (\Sigma\Omega E_T, \Sigma\Omega E) & \xrightarrow{\Sigma\Omega T(f)} & (\Sigma\Omega E_{0T}, \Sigma\Omega E_0) \\ \downarrow & & \downarrow \\ (E_T, E) & \xrightarrow[T(f)]{} & (E_{0,T}, E_0) \end{array}$$

with evaluation along the vertical edges. The constructions in (7.2.1) yield the following ladder of exact sequences,

$$\begin{array}{ccccccccc} 0 & \longrightarrow & \Sigma R(\Omega\xi) & \longrightarrow & \Sigma N(\Omega\xi) & \longrightarrow & \Sigma M(\Omega\xi) & \longrightarrow & 0 \\ & & \uparrow & & \uparrow & & \uparrow & & \\ 0 & \longrightarrow & R(\xi) & \longrightarrow & N(\xi) & \longrightarrow & M(\xi) & \longrightarrow & 0 \ , \end{array}$$

and (ii) follows by use of the structure theorems. □

We work another example with both the coproduct theorem and the fundamental sequence in play. Fix n and consider the sequence of maps on display below (suppressed coefficients are $Z/2$)

$$K(Z,n+k) \xrightarrow[Sq^n]{} K(2n+k) \xrightarrow[\binom{Sq^1}{Sq^2}]{} K(2n+k+1, 2n+k+2)$$

$$\xrightarrow[(Sq^3,Sq^2)]{} K(2n+k+4) \ .$$

If n is even, we must take $k=0$ in order to have a syzygy, and for $n \equiv 3(4)$, $k=0$ and 1, while for $n \equiv 1(4)$ we may take $k \geq 0$. Thus we study

$$\varphi_{1,1}(Sq^n)$$

when this composition is defined.

With $k=0$ and W_n denoting the homotopy fiber of Sq^n, regarded as a loop space, the cohomology is given by

$$H^*(W_n) = \Lambda(Sq(I)u_n) \otimes U(\Omega Q)$$

when I runs over admissible monomials not ending in 1 and $exI < n$. The module $Q \subset F_{2n}$ is spanned by admissible monomials with at least one odd entry or ending with 2. We can regard ΩQ as a submodule of F_{2n-1} and the map

$$G_n \to \Omega Q \to 0$$

from the fundamental sequence is the restriction to the fiber. The first two elements of ΩQ are

$$Sq^1 b_{2n-1} \ , \quad Sq^2 b_{2n-1} \ .$$

We shall write e_{2n}, e_{2n+1} for their preimages in G_n and make certain choices.

Now restrict n to be even. By the coproduct theorem, we may choose e_{2n+1} to be primitive and e_{2n} to have middle term

$$p^*u_n \otimes p^*u_n \ .$$

The element

$$g = Sq^3 e_{2n} + Sq^2 e_{2n+1} + \text{ terms from } im\, p^*$$

maps to 0 in ΩQ and we wish to find the unknowns from

$$p^* : \Lambda(Sq(I)u_n) \to G_n \ .$$

Expanding g by the coproduct yields middle terms

$$p^*u_n \otimes p^*Sq^3u_n + p^*Sq^3u_n \otimes p^*u_n + \text{ coproduct on } im\, p^* \ .$$

Since p^* is monic on the exterior algebra, the unknown terms must be of the form

$$u_n \cup Sq^3u_n + \theta u_n$$

where θ is a Steenrod operation of degree $n+3$.

We can determine θ by looping W_n. Here

$$\Omega W_n \simeq K(Z, n-1) \times K(Z/2, 2n-2) .$$

By the coproduct theorem, we may choose the splitting so that the coproduct on b_{2n-2} has middle term

$$\Omega p^* u_{n-1} \otimes \Omega p^* u_{n-1} .$$

Now

$$\sigma^* g = Sq^3 \sigma^* e_{2n} + Sq^2 \sigma^* e_{2n+1} + \theta u_{n-1} ;$$

moreover, $\sigma^* : PH^{2n+3}(K(Z,n)) \to PH^{2n+2}(K(Z,n-1))$ is an isomorphism–epic in general and monic as there are no contributions from admissible monomials of excess n, in this dimension. In terms of the splitting for ΩW_n, we have the following equations for the suspensions of the e's,

$$\sigma^* e_{2n} = Sq^1 b_{2n-2} + \theta_1 u_{n-1} ,$$
$$e^* e_{2n+1} = Sq^2 b_{2n-2} + u_{n-1} \cup Sq^2 u_{n-1} + \theta_2 u_{n-1}$$

where θ_1, θ_2 are elements of the Steenrod algebra. We may use either one to alter the tetherings for e_{2n}, e_{2n+1} without changing the coproduct formulas, because σ^* is epic. So we can eliminate these terms from further consideration.

Now we obtain

$$\sigma^* g = \Omega p^* Sq^2(u_{n-1} \cup Sq^2 u_{n-1}) = \Omega p^* \theta u_{n-1} .$$

Since

$$Sq^2(u_{n-1} \cup Sq^2 u_{n-1}) = Sq^{n+1} Sq^2 u_{n-1}$$

and σ^* is monic, we find that

$$\theta = Sq^{n+1} Sq^2 .$$

□

We restate the result as

$$\varphi_{1,1}(Sq^n u_n) = u_n \cup Sq^3 u_n + Sq^{n+1} Sq^2 u_n \text{ in } H^{2n+3} / Sq^3 H^{2n} + Sq^2 H^{2n+1} .$$

Notice that possible changes in tetherings for e_{2n}, e_{2n+1} are absorbed in the indeterminancy and $\varphi_{1,1}$ is not affected by variation of tetherings.

Another way to report the information is that there exist secondary operations Ψ_1, Ψ_2 based on

$$Sq^1 Sq^n , \quad Sq^2 Sq^n$$

such that for n-dimensional mod 2 reductions of integral classes with cup square equal to 0,

$$Sq^3 \Psi_1(x) + Sq^2 \Psi_2(x) = x \cup Sq^3 x + Sq^{n+1} Sq^2 x$$

in

$$H^{2n+3}/Sq^3Sq^1H^{2n-1}.$$

The refinement of the indeterminancy is at the price of fixing Ψ_1 and Ψ_2.

A third way to report the information is in the form of a conditional relation. For x as above with $x^2 = 0$, we have

$$x \cup Sq^3x + Sq^{n+1}Sq^2x = Sq^3y_1 + Sq^2y_2$$

for some classes y_1, y_2.

The situation for n odd, $k = 0$ is similar. Now both e_{2n} and e_{2n+1} may be chosen to be primitives. The result is

$$\varphi_{1,1}(Sq^nu_n) \equiv Sq^{n+1}Sq^2u_n \text{ modulo Ind } (\varphi_{1,1}).$$

Remark. For $n = 3$, we may use an Adem argument with the relation above to prove that the composition

$$\eta_3 \circ \nu_4 : S^7 \longrightarrow S^3$$

is essential. We leave this as an exercise as well as an investigation of the cases $k > 0$. Results first obtained by Mahowald [**66**] may be worked out this way. The separation of the values of n into congruence classes 1 or 3 mod 4 is not an artifact of the algebra. The single suspension of $\eta_3 \circ \nu_4$ is essential, but its double suspension is 0.

7.2.4. Unstable Adams resolutions. Massey-Peterson fibrations hold the key to extending the basic ideas, used initially by Adams for stable homotopy theory [**1**], to spaces, albeit with highly restricted cohomology of the form $U(M)$. A general approach was discovered by Bousfield and Kan which (rightly) puts the Massey-Peterson construction in less prominence insofar as general results are concerned. Both points of view bring towers of fibrations as a tool for resolving homotopy types. They augment the basic Moore-Postnikov approach by tieing the tower to homological algebra on the homology of the space at hand. In this section we describe the basic features of unstable Adams resolutions for the Massey-Peterson case. We refer to [**74**], [**39**], [**42**] for the many details not supplied here. The main purpose for including this section is to prove a recognition principle for the Lie group G_2. This result is mentioned in [**7**] so a proof should appear somewhere. The result is the following. *If X is a simply connected space with mod 2 cohomology ring isomorphic to*

$$Z/2[x_3]/(x_3^4) \otimes \Lambda(x_5)$$

then there is a homotopy equivalence of X with G_2 after localizing at 2. If X is a 2-local H-space, the hypothesis of simple connectivity is not required. The proof of this result is a routine consequence of an unstable Adams resolution for G_2, which in turn is a routine and not lengthy calculation. It will be displayed in due course.

We turn to the resolutions. Suppose X is a space with mod p cohomology of the form $U(M)$. Suppose we have a projective resolution of M as an unstable module,

$$P_* \to M .$$

We can make a syzygy

$$K(P_0) \leftarrow K(P_1) \leftarrow \cdots \leftarrow K(P_s) \xleftarrow{f_s} K(P_{s+1}) \leftarrow \cdots$$

where $f_s = K(d_s)$ for $d_s : P_{s+1} \to P_s$. The theory of unstable Adams resolutions may be summarized by saying that any finite initial segment of this syzygy is admissible, in the sense of Chapter 4, Section 3. The admissibility is a consequence of the fact that the tower consists of Massey-Peterson fibrations and the space X forces a splitting of their fundamental sequences.

Our application to G_2 will illustrate these remarks, so we turn to it now. First of all, the cohomology ring forces the Steenrod action. Since

$$Sq^3 x_3 = x_3^2 \neq 0 \text{ and } Sq^3 = Sq^1 Sq^2 ,$$

we must have $Sq^2 x_3 = x_5$ and $Sq^1 x_5 = x_3^2$. For dimensional reasons, the only other possible action involves

$$Sq^4 Sq^2 x_3 ,$$

and the possible values are 0 or x_3^3. But, by the above information for Sq^2, we have

$$Sq^2 x_3^3 = x_5 \cdot x_3^2 \neq 0 .$$

Thus if

$$Sq^4 Sq^2 x_3 = x_3^3 ,$$

then

$$Sq^2 Sq^4 Sq^2 x_3 = x_5 \cdot x_3^2 \neq 0 .$$

But

$$Sq^2 Sq^4 = Sq^6 + Sq^5 Sq^1 \text{ and } Sq^5 Sq^3 = 0$$

so the equation cannot hold.

Write M for the unstable module

$$M = \text{ span } \{x_3, Sq^2 x_3, Sq^3 x_3\} .$$

Then the cohomology of G_2 is $U(M)$ and likewise for X.

We construct a resolution for M where we take advantage of the Sq^1 information to use the modules F_n' in place of F_n, where possible. This means that we start with F_3' instead of F_3.

The kernel of the first map

$$F_3' \to M$$

is all of F_3' above dimension 6, which we can write as

$$\{Sq^4Sq^2u_3,\ Sq^5Sq^2u_3,\ Sq^6Sq^3u_3,\ Sq^8Sq^4Sq^2u_3\ ,$$
$$Sq^{16}Sq^8Sq^2u_3,\ \cdots\} \cup \{\text{ iterates of } \lambda \text{ on these elements}\}$$

where $\lambda x = Sq^n x, n = |x|$. This module is evidently a cyclic $\mathcal{A}$-module on the generator in dimension 9. So we have an initial segment of a minimal resolution given by

$$M \leftarrow F_3' \xleftarrow{d_0} F_9,\ d_0(b_9) = Sq^4Sq^2u_3\ .$$

For the rest of the calculation, we focus our attention only on an initial segment of degrees sufficient to establish the cohomology structure of each space in the tower in dimensions ≤ 14, as this is the dimension of G_2 or of a homology approximation to X. For d_0 this means dimensions ≤ 15. As a practical matter I recommend calculating a bit farther than necessary. We skip to a description of the result.

The following elements are independent elements of $\ker d_0$

$$\{Sq^2b_9,\ Sq^4b_9,\ Sq^4Sq^2Sq^1b_9\}\ .$$

Thus we may continue our construction with

$$M \leftarrow F_3' \xleftarrow{d_0} F_9 \xleftarrow{d_1} F_{11} \oplus F_{13} \oplus F_{16}$$

with

$$d_1 = \begin{pmatrix} Sq^2 \\ Sq^4 \\ Sq^4Sq^2Sq^1 \end{pmatrix}\ .$$

Before going on, let us pause to write the portion of the tower implied by our calculation

$$\begin{array}{ccc} E_1 & \xrightarrow{\tilde{d}_1} & K(10,12,15) \\ \big\downarrow & & \\ K(Z,3) & \longrightarrow & K(9)\ . \end{array}$$

The space E_1 is a Massey-Peterson fibration with cohomology

$$U(M) \otimes U(\Omega \ker d_0)\ .$$

We have G_2 mapping to $K(Z,3)$ so as to realize the map $F_3' \to M$ and consequently the fundamental sequence for E_1 splits. Hence, the description above is as algebras over the Steenrod algebra (but not necessarily as Hopf algebras and we shall return to this point later). The splitting means that we may choose the colifting $\tilde{d_1}$ so that the composition

$$\Omega P_2 \xrightarrow{\tilde{d_1}^*} G_1 \to \Omega \ker d_0 \hookrightarrow \Omega P_1$$

is Ωd_1. Moreover, we have the Massey-Peterson conditions holding for $\tilde{d_1}$, so we can go to the next stage

$$\begin{array}{ccc} E_2 & & \\ \downarrow & & \\ E_1 & \longrightarrow & K(10,12,15) \\ \downarrow & & \\ K(Z,3) & \longrightarrow & K(9) \end{array}$$

with $H^*(E_2) \cong U(M) \otimes U(\Omega \ker \Omega d_1)$ again as algebras over the Steenrod algebra. Moreover, we have

$$\Omega \ker \Omega d_1 = \Omega^2 \ker d_1$$

by the lemma on p. 16 of [**39**], (this is an elementary Ω_1 calculation).

Hand calculation produces d_2,

$$F_{11} \oplus F_{13} \oplus F_{16} \xleftarrow{d_2} F_{14} \oplus F_{17} \oplus F_{17} \oplus F_{18}$$

with matrix representation

$$\begin{bmatrix} Sq^3 & 0 & 0 \\ Sq^6 & Sq^4 + Sq^3Sq^1 & 0 \\ Sq^4Sq^2 & 0 & Sq^1 \\ Sq^4Sq^2Sq^1 & Sq^4Sq^1 & Sq^2 \end{bmatrix}.$$

Now look at the top left corner entry Sq^3. Since $Sq^1Sq^3 = 0$, an infinite tower results from this element in $\ker \tilde{d_1}^*$. In terms of our towers, we have

$$\begin{array}{ccc} E_2 & \xrightarrow{e} & K(Z/2,12) \\ {\scriptstyle p_1}\downarrow & & \\ E_1 & & \end{array}$$

restricting to Sq^3b_9 in the fiber and we would like to know whether e is the mod 2 reduction of an integral class. But the value of any Bockstein on e lies in the image of p_1^* and only $U(M)$ is mapped non-zero by p_1^*. As this algebra is 0 in dimension 12, we may so represent e. We replace F_{14} by F'_{14} in the source of d_2.

Rationally,

$$G_2 \simeq S^3 \times S^{11}$$

and we are seeing how the infinite summand of π_{11} arises in the tower.

We make a hand calculation of the modified d_2 and the result is

$$F'_{14} \oplus F_{17} \oplus F_{17} \oplus F_{18} \xleftarrow{d_3} F_{18}$$

with $d_3 = [Sq^4, 0, Sq^1, 0]$.

We have now produced a tower for G_2 and we display the dimensions of its k-invariants

$$\begin{array}{ccc} E_3 & \longrightarrow & K(15) \\ \downarrow & & \\ E_2 & \longrightarrow & K(12^*, 15, 15, 16) \\ \downarrow & & \\ E_1 & \longrightarrow & K(10, 12, 15) \\ \downarrow & & \\ K(Z,3) & \longrightarrow & K(9) \end{array}$$

and the map $G_2 \to E_3$ induces an isomorphism in cohomology in dimensions ≤ 14. The asterisk indicates Z in place of $Z/2$ in dim. 12.

We map $X \to K(Z,3)$ to represent X_3. We have seen that $Sq^4Sq^2x_3 = 0$, so our map lifts to E_1. It lifts to E_3 for dimensional reasons, and likewise to G_2. Since the lifted map works on cohomology according to how X was mapped to $K(Z,3)$, we have obtained a map $X \to G_2$ inducing an isomorphism on mod 2 cohomology. Thus the 2-localization are homotopy equivalent. □

Remark. We look at what the coproduct theorem says about E_1 when we regard it as a loop space obtained from looping the homotopy fiber of

$$K(Z,4) \xrightarrow[Sq^4Sq^2]{} K(Z/2, 10) .$$

The result is that the 10-dimensional class is not primitive and has middle term

$$Sq^2u_3 \otimes Sq^2u_3 .$$

This does not contradict that G_2 is an H-space. The H-structure on E may be changed by

$$E_1 \wedge E_1 \to K(Z_2, 8)$$

representing $u_3 \otimes Sq^2 u_3$ and the result gives a primitive 10-dimensional class and does not change the others, which were primitive to begin with.

Moreover, the displayed map is not the first k-invariant for BG_2. We have

$$H^* BG_2 = Z/2[x_4, Sq^2 x_4, Sq^3 x_4]$$

and $Sq^4 Sq^2 x_4 = x_4 \cdot Sq^2 x_4$.

Remark. The Massey-Peterson spectral sequence calculates a portion of the 2-primary part of the homotopy of G_2 in our displayed range of stems ≤ 15. There are no differentials and $E_2 = E_\infty$. Extension issues can be settled in terms of the displayed information as well, because there are no jumps in the Adams filtration. We recover well-known information, the non-zero groups are (2 primary only)

	gen.	Adams filtration of generator
$\pi_3 = Z$	inc. of S^3	0
$\pi_8 = Z/2$	a	1
$\pi_9 = Z/2$	$a \circ \eta_8$	2
$\pi_{11} = Z/2 + Z$	$a \circ \nu_8, b$	2,3
$\pi_{14} = Z/8 + Z/2$	c, d	2,3
$\pi_{15} = Z/2$	$c \circ \eta_{14}$	3

Moreover, $4c = b \circ \nu_{11}$ in π_{14}.

7.3. Products

In this section, we develop operations which incorporate Cartan formula information into the defining relations. We shall discuss twisted coliftings, twisted secondary operations and Cartan formulas for secondary operations. In the literature, these topics are developed in papers by Gitler and Stasheff, Mahowald, Meyer, Milgram, McClendon, Peterson, Ravenel, Thomas, among others. Papers by Massey, Peterson and Stein made calculations which set the stage for this part of the subject.

7.3.1. Half-smash product construction. We review a construction introduced in Chapter 3. It will be used in each of the topics treated here.

If A is a space, A^+ is A with a disjoint basepoint. We shall make this step even if A happens to be pointed. There is a canonical map (counit)

$$A^+ \xrightarrow{i} S^0 = \{\pm 1\}$$

sending the disjoint basepoint to $+1$ and A to -1. If A is already pointed, then there is a canonical map (unit)

$$S^0 \xrightarrow{d} A^+$$

sending $+1$ to the disjoint basepoint and -1 to the basepoint of A. The composition of the unit with the counit, id, is the identity map on S^0.

By definition

$$A \ltimes B = A^+ \wedge B \ .$$

In the sequel, we work in the pointed category $\mathcal{K}_*$ unless otherwise stipulated. Note that B is a canonical retract of $A \ltimes B$ via $i \wedge 1$. Moreover,

$$B = S^0 \wedge B \xrightarrow{d \wedge 1} A \ltimes B \to A \wedge B$$

is a cofibration sequence.

Given a map

$$f : A \ltimes B \to C$$

we construct the semi-evaluation map

$$\hat{f} : A \ltimes PB \to PC$$

given by

$$\hat{f}(a, \lambda)(s) = f(a, \lambda(s)) \ .$$

Observe that if $s = 0$, then $\hat{f}(a, \lambda)(0) = f(a, *) = *$.

The principal properties of the semi-evaluation map are

$$\text{(i)} \qquad \hat{f}(a, \sigma + \lambda) = \hat{f}(a, \sigma) + \hat{f}(a, \lambda)$$

for σ in ΩB and λ in PB and (ii), the following diagram commutes

$$\begin{array}{ccc} \Sigma(A \ltimes \Omega B) & \xrightarrow{\Sigma \hat{f}} & \Sigma \Omega C \\ \downarrow & & \downarrow \\ A \ltimes B & \xrightarrow[f]{} & C \end{array}$$

where the vertical map on the left is given by

$$(t, a, \lambda) \to (a, \lambda(t))$$

and the vertical map on the right is evaluation. In terms of cohomology classes, (ii) says that if

$$f^*(c) = \Sigma a_i \otimes b_i \ ,$$

then

$$\hat{f}^*(\sigma^* c) = \Sigma a_i \otimes \sigma^* b_i \ .$$

7.3.2. Twisted coliftings. We have used the colifting construction to represent cohomology of certain fiber spaces. Here we modify that construction to achieve a geometric means of representing cohomology in more general situations. To introduce this modification, we first look at an example which inspired this theory.

Consider the fiber square

$$\begin{array}{ccc} E & \xrightarrow{h} & PK(Z,n) \\ {\scriptstyle p}\downarrow & & \downarrow \\ BSO(n) & \xrightarrow[\chi]{} & K(Z,n) \end{array}$$

where χ is the Euler class. We shall calculate the mod 2 cohomology of E using the Serre spectral sequence. The mod 2 reduction of χ is the Stiefel-Whitney class w_n and we shall use the Wu formula

$$Sq^k w_n = \sum_{t=0}^{k} \binom{n-t-1}{k-t} w_{n+k-t} w_t \ .$$

Here the usual conventions for binomial coeffients are in force with the single exception that

$$\binom{-1}{0} = 1 \ .$$

We have $E_2 = E_n$ and the first non-trivial differentials are

$$d_n(1 \otimes u_{n-1}) = w_n \otimes 1 \ ,$$
$$d_n(w_2 \otimes u_{n-1}) = w_2 \cdot w_n \otimes 1 \ .$$

Hence $\{1 \otimes Sq^2 u_{n-1}\}$ is a permanent cycle in $E_{n+2}^{0,n+1}$. Continuing in this manner [**92**] we obtain

$$H^*(E) \cong H^*(BSO(n-1)) \otimes Z/2[SqIu_{n-1}]$$

where I ranges over admissible monomials of positive degree, not terminating in 1, and having excess $< n-1$ with the single exception of $I = (n-1)$.

The relation

$$Sq^2 w_n = w_2 \cdot w_n$$

gives a sequence with homotopy

$$BSO(n) \xrightarrow{1 \ltimes \chi} BSO(n) \ltimes K(Z,n) \xrightarrow{\alpha} K(Z/2, n+2)$$

where $\alpha^*(b_{n+2}) = w_2 \otimes u_n + 1 \otimes Sq^2 u_n$.

We shall use the null-homotopic composition of $\alpha \circ (1 \ltimes \chi)$ to produce a map

$$\tilde{\alpha} : E \to K(Z/2, n+1)$$

such that the restriction to the fiber gives $Sq^2 u_{n-1}$.

We turn to the generic situation. Suppose we have a fiber square

$$\begin{array}{ccc} W & \xrightarrow{h} & PB \\ {\scriptstyle p}\downarrow & & \downarrow \\ A & \xrightarrow[\alpha]{} & B \end{array}$$

and a null-homotopic composition

$$A \xrightarrow{1\ltimes\alpha} A \ltimes B \xrightarrow{\beta} C\,.$$

Note that $(1 \ltimes \alpha)(a) = (a, \alpha(a))$ and we have suppressed the diagonal map from the notation. Let β_1 denote the composition

$$B = S^0 \wedge B \xrightarrow{d\wedge 1} A \ltimes B \xrightarrow{\beta} C\,.$$

Let N be a null-homotopy for the composition $\beta \circ (1 \ltimes \alpha)$. Then there is a map (called a <u>twisted colifting</u>)

$$\tilde{\beta} : W \to \Omega C$$

satisfying

(i)

$$\begin{array}{ccc} \Omega B & \xrightarrow{j} & W \\ {\scriptstyle \Omega\beta_1}\downarrow & & \downarrow{\scriptstyle \tilde{\beta}} \\ \Omega C & = & \Omega C \end{array}$$

up to homotopy,

(ii) The principal action satisfies

$$\begin{array}{ccccc} \Omega B \times W & \xrightarrow{\mu} & W & \xrightarrow{\tilde{\beta}} & \Omega C \\ {\scriptstyle 1\times\Delta}\downarrow & & & & \uparrow \\ \Omega B \times W \times W & & & & {\scriptstyle +} \\ {\scriptstyle T\times 1}\downarrow & & & & \\ W \times \Omega B \times W & \xrightarrow[p\ltimes 1\times 1]{} & A \ltimes \Omega B \times W & \xrightarrow[\hat{\beta}\times\tilde{\beta}]{} & \Omega C \times \Omega C \end{array}$$

commutes up to homotopy,

(iii) (a Peterson-Stein formula) Given a homotopy commutative diagram

$$\begin{array}{ccccccc}
X & \xrightarrow{f} & Y & \xrightarrow{j} & T_f & \xrightarrow{q} & \Sigma X \\
\downarrow{\scriptstyle L} & & \downarrow{\scriptstyle \varepsilon} & & \downarrow{\scriptstyle \zeta} & & \downarrow{\scriptstyle E} \\
W & \xrightarrow[p]{} & A & \xrightarrow[1\ltimes\alpha]{} & A \ltimes B & \xrightarrow[\beta]{} & C \\
\downarrow{\scriptstyle \tilde{\beta}} & & & & & & \\
\Omega C & & & & & &
\end{array}$$

then $[\![\tilde{\beta} \circ L]\!] = [\![E^\natural]\!]$ in

$$[X, \Omega C]/im\, f^{\#} + im\, \hat{\beta}_{\#}$$

where $\hat{\beta}_{\#} : [X, \Omega B] \to [X, \Omega C]$ is given by

$$\delta \to \hat{\beta} \circ (\varepsilon f \ltimes \delta) \ ,$$

(iv) If $(\beta, 1\ltimes\alpha, N_1)$ and $(\beta, 1\ltimes\alpha, N_2)$ are two sequences with homotopy, then the corresponding coliftings differ by $\delta \circ p$ where $\delta = \{N_1, \tau \circ N_2\}$. Moreover, the homotopy class of the colifting is an invariant of the homotopy class of the sequence with homotopy. In particular, the value of $[\![\tilde{\beta} \circ L]\!]$ in (iii) is independent of the choice of contracting homotopy.

Proof. We construct $\tilde{\beta}$ from the following pull-back data

$$\begin{array}{ccccc}
W & = & W & = & W \\
\downarrow{\scriptstyle p} & & \downarrow & & \downarrow{\scriptstyle p\ltimes h} \\
A & \xrightarrow[1\ltimes\alpha]{} & A \ltimes B & \longleftarrow & A \ltimes PB \\
\downarrow{\scriptstyle N_\tau^\natural} & & \downarrow{\scriptstyle \beta} & & \downarrow{\scriptstyle \hat{\beta}} \\
P_1 C & \longrightarrow & C & \longleftarrow & P_0 C
\end{array}$$

where the unnamed vertical map is $(1 \ltimes \alpha) \circ p$, and the other unnamed maps are evaluation maps. In terms of elements,

$$\tilde{\beta}(x, \lambda)(s) = \begin{Bmatrix} \beta(x, \lambda(2s)) & 0 \le s \le \frac{1}{2} \\ N(2 - 2s, x) & \frac{1}{2} \le s \le 1 \end{Bmatrix} \ .$$

Thus for (i) we have,

$$\tilde{\beta} \circ j(\sigma) = \tilde{\beta}(*, \sigma) \simeq \beta_1 \circ \sigma \ .$$

Part (ii) is immediate from the semi-evaluation formula for $\hat{\beta}$.

For part (iii), we observe that $A \ltimes PB$ is contractible and

$$(1 \ltimes \alpha) \circ \varepsilon f \sim *$$

if and only if $\alpha\varepsilon f \sim *$.

Thus (iii) follows from Cor. 3.4.1 and the pull-back data for $\tilde{\beta}$.

Finally, (iv) follows as our construction of $\tilde{\beta}$ is really a special case of the general colifting construction. $\square$

We shall give some representative applications of the twisted colifting. Underlying many calculations is the observation that we may identify the pull-back of $p : W \to A$ over itself with the principal action,

$$\begin{array}{ccc} \Omega B \times W & \xrightarrow{u} & W \\ \Big\downarrow{\scriptstyle r} & & \Big\downarrow{\scriptstyle p} \\ W & \xrightarrow[p]{} & A \end{array} \qquad r = \text{ projection on second factor.}$$

This observation is combined with property (ii) for twisted coliftings.

For our first application, consider the Whitehead product, [**90**]

$$[g, p_n] : S^{2n+2} \to CP^n$$

where $p_n : S^{2n+1} \to CP^n$ is the standard fiber map and $g : S^2 \to CP^n$ is inclusion of the bottom cell. We prove that for n even, $[g, p_n] \neq 0$. To begin we have the following homotopy commutative diagram

$$\begin{array}{ccccc} & & K(Z, 2n+1) & & \\ & \overset{-1}{\nearrow} & \Big\downarrow & & \\ S^{2n+1} & & E & & \\ \Big\downarrow{\scriptstyle p_n} & \overset{f}{\nearrow} & \Big\downarrow & & \\ CP^n & \xrightarrow[y]{} & K(Z, 2) & \xrightarrow[b_2^{n+1}]{} & K(Z, 2n+2) \end{array}$$

where -1 is the colifting determined by $y^{n+1} = 0$ in the cohomology of CP^n and the sign involved in comparing with the lifting f has been incorporated into this map. Since n is even, we have a set-up for a twisted colifting,

$$K(Z, 2) \xrightarrow[1 \ltimes b_2^{n+1}]{} K(Z, 2) \ltimes K(Z, 2n+2) \xrightarrow{\alpha} K(Z/2, 2n+4)$$

with $\alpha^*(b_{m+4}) = b_2 \otimes u_{2n+2} + 1 \otimes Sq^2 u_{2n+2}$. The composition is null because

$$Sq^2 b_2^{n+1} = \binom{n+1}{1} b_2^{n+2} = b_2^{n+2}\,.$$

We may now form the following homotopy commutative diagram

$$\begin{array}{ccccccc}
S^{2n+2} & \xrightarrow{\ w\ } & S^{2n+1} \vee S^2 & \xrightarrow{(-1, f\circ g)} & K_{2n+1} \times E & \longrightarrow & K(Z/2, 2n+3) \\
\Vert & & \Big\downarrow{\scriptstyle p_n \vee g} & & \Big\downarrow{\scriptstyle \mu} & & \Vert \\
S^{2n+2} & \xrightarrow[{[p_n, g]}]{} & CP^n & \xrightarrow[f]{} & E & \xrightarrow[\tilde{\alpha}]{} & K(Z/2, 2n+3)\,.
\end{array}$$

Here, w is the universal Whitehead product and the top row is the set-up for the functional cup product because

$$b_{2n+3} \to Sq^2 b_{2n+1} \otimes 1 + b_{2n+1} \otimes p^* b_2 + 1 \otimes \tilde{\alpha}$$

and the middle term detects w. The total indeterminancy is 0 as both maps

$$\pi_{2n+2}\Omega(K_{2n+1} \times E) \to \pi_{2n+2}K(Z/2, 2n+2)$$

and

$$[CP^n, K(Z/2, 2n+2)] \to \pi_{2n+2}K(Z/2, 2n+2)$$

have 0 image. The assertion that $[p_n, g] \neq 0$ now follows by naturality. Moreover, the homotopy sequence for a fibration yields that

$$[p_n, g] = p_n \circ \eta_{2n+1} \qquad n \text{ even.}$$

□

For n odd, the calculation

$$Sq^2 b_2^n = b_2^{n+1}$$

implies that the composition of p_n with the pinch map from CP^n to the top cell S^{2n} is η_{2n}. It follows that

$$[p_n, g] = 0 \quad n \text{ odd}$$

since $\eta_{2n} \circ \eta_{2n+1} \neq 0$. Of course, the colifting construction for this case would give 0, but not the conclusion, since a priori there could be other ways to detect the Whitehead product. For other results of this type, see Barratt, James and Stein [**9**].

Remark. The twisted colifting $\tilde{\alpha}$ detects $p_n \circ \eta_{2n+1}$ for even values of n. To see this we use the Peterson-Stein property (iii) in conjunction with the cofibration sequence for the composition $p \circ \eta$,

$$CP^n \cup e^{2n+3} \to CP^{n+1} \to S^{2n+2} \cup_\eta e^{2n+4}\,.$$

To obtain the initial set-up, we have the following general principle. A composition

$$X \xrightarrow{g} D \xrightarrow{\alpha} E$$

is null-homotopic if and only if

$$(1 \ltimes \alpha) \circ g$$

is null-homotopic. This fact is evident when we write the latter compositon without the suppression of diagonal maps,

$$(1 \ltimes \alpha) \circ g = (g \ltimes d \circ g) \circ \bar{d}$$

where

$$\bar{d} : X \to X \ltimes X$$

is the composition of the diagonal and the natural projection.

To set up (iii), we have the following homotopy commutative square

$$\begin{array}{ccc} CP^{n+1} & \longrightarrow & S^{2n+2} \cup e^{2n+4} \\ \downarrow & & \downarrow \\ K(Z,2) & \xrightarrow[1 \ltimes b_2^{n+1}]{} & K(Z,2) \ltimes K(Z,2n+2) \end{array}$$

where the left vertical map represents a generator of H^2 and the right vertical map induces an isomorphism on H^{2n+2}, both cohomologies with integral coefficients. The remaining steps in the argument are left for the reader. For odd values of n, an ordinary colifting detects the composition.

Other examples where the same argument may be made are

$$S^3 \xrightarrow{\eta} S^2 \to RP^2$$

and the twisted colifting based on

$$K(Z/2,1) \xrightarrow[1 \ltimes b_1^3]{} K(Z/2,1) \ltimes K(Z/2,3) \xrightarrow[\alpha]{} K(Z/2,5)$$

with

$$\alpha^*(b_5) = b_1^3 \otimes b_3 + 1 \otimes Sq^2 b_3 \ .$$

For odd primes, we have an example,

$$S^{2p} \xrightarrow{\alpha_1} S^3 \xrightarrow{\eta} S^2 \ ,$$

and

$$K(Z,2) \xrightarrow[1 \ltimes u_2^2]{} K(Z,2) \ltimes K(Z/p,4) \xrightarrow[\alpha]{} K(Z/p,2p+2)$$

with

$$\alpha^*(b_{2p+2}) = -2u_2^{p-1} \otimes u_4 + 1 \otimes \mathcal{P}^1 u_4 \ .$$

The next example treats a twisted colifting in the role of a k-invariant. We have the bundle

$$S^{n-1} \xrightarrow{k} BSO(n-1) \xrightarrow{\pi} BSO(n)$$

which classifies orientable sphere bundles. The first k-invariant of π is the Euler class χ, and there is the following diagram expressing the fact that χ represents the transgression of a generator from the fiber. We abbreviate $BSO(k)$ to B_k.

$$\begin{array}{ccccccc} S^{n-1} & \xrightarrow{k} & B_{n-1} & \longrightarrow & B_{n-1} \cup e^n & \longrightarrow & S^n \\ {\scriptstyle -1}\big\downarrow & & {\scriptstyle r}\big\downarrow & \overset{\pi}{\searrow} & \big\downarrow & & \big\downarrow{\scriptstyle 1} \\ K(Z, n-1) & \xrightarrow[j]{} & E & \xrightarrow[p]{} & B_n & \xrightarrow[\chi]{} & K(Z, n)\ . \end{array}$$

Suppose we have an orientable sphere bundle and its classifying map

$$\begin{array}{ccc} V & \xrightarrow{\lambda} & B_{n-1} \\ {\scriptstyle \rho}\big\downarrow & & \big\downarrow{\scriptstyle \pi} \\ X & \xrightarrow[\xi]{} & B_n \end{array}$$

and suppose the Euler class vanishes, $\chi \circ \xi \sim *$. Then we can lift ξ to E and compare the two ways around the following square:

$$\begin{array}{ccc} V & \xrightarrow{\lambda} & B_{n-1} \\ {\scriptstyle \rho}\big\downarrow & & \big\downarrow{\scriptstyle r} \\ X & \xrightarrow[f]{} & E\ . \end{array}$$

Define $e : V \to K(Z, n-1)$ by

$$e(x) = \{hr\lambda(x), H_\tau \circ \rho(x)\}$$

where H is a contracting homotopy for $\chi \circ \xi$ and

$$h : E \to PK(Z, n)$$

is from the fiber square for E.

Then the following diagrams commute up to homotopy,

$$\begin{array}{ccc} V & \xrightarrow{(e, r\circ\lambda)} & K(Z, n-1) \times E \\ {\scriptstyle \rho}\big\downarrow & & \big\downarrow{\scriptstyle \mu} \\ X & \xrightarrow[f]{} & E\ , \end{array}$$

and

$$\begin{array}{ccc} V & \xrightarrow{e} & K(Z, n-1) \\ {\scriptstyle i}\uparrow & & \| \\ S^{n-1} & \xrightarrow[-1]{} & K(Z, n-1) \end{array} \qquad i = \text{ inclusion of the fiber.}$$

The first fact is an explicit rendering of the principal action while the second follows by direct verification with the formula for e,

$$\begin{aligned} e \circ i(x) &= \{hr\lambda i(x), H_\tau \circ \rho i(x)\} \\ &\simeq \{hrk(x)\} \\ &= \{hj(-1)(x)\} \\ &= \{(-1)x\} \ . \end{aligned}$$

In this material we are following Thomas [**116**], section 5.

We have used twisted coliftings to represent classes in $H^{n+1}(E)$ which restrict to $Sq^2 u_{n-1}$ in the fiber. If, in addition, the composition

$$B_{n-1} \xrightarrow{r} E \xrightarrow{\tilde{\alpha}} K(Z/2, n+1)$$

is null-homotopic, then $\tilde{\alpha}$ represents the next k-invariant. In our case, this fact follows easily from our earlier calculations. The general situation (for k-invariants corresponding to the initial stable homotopy groups of the fiber) is discussed in two papers of Thomas [**113**], [**114**]. Moreover, we have the following calculation for this k-invariant, using property (ii) for twisted coliftings,

$$\rho^* f^* \tilde{\alpha}^*(b_{n+1}) = \rho^* w_2(\xi) \cup e + Sq^2 e \ .$$

in $H^*(V; Z/2)$.

Next, we look at the Thom complex over E in order to gain more information about $\tilde{\alpha}$. We are following Mahowald and Peterson [**68**] in this development. Write T for the Thom complex over E obtained from the pull-back of the universal n-plane bundle over B_n. Write U for the Thom class. Then

$$Sq^n U = p^* w_n \cup U = 0 \ .$$

Thus the secondary operation Θ_n, based on $Sq^2 Sq^n$, and discussed in Chapter 5, section 1, is defined on U. We evaluate the Θ_n of Proposition 5.1.1 on U. The result is, for twisted coliftings $\tilde{\alpha}$ with $\tilde{\alpha} \circ r \sim *$,

$$\Theta_n(U) = [\![U \cdot (\tilde{\alpha} + p^*(w_2 \cdot w_{n-1})]\!] \ .$$

Proof. We have the sphere bundle over E obtained by pull-back from π,

$$\begin{array}{ccc} V & \xrightarrow{\lambda} & B_{n-1} \\ {\scriptstyle\rho}\downarrow & & \downarrow{\scriptstyle\pi} \\ E & \xrightarrow[p]{} & B_n \end{array} .$$

From our general discussion we have

$$e : V \to K(Z, n-1)$$

such that the following diagram commutes up to homotopy:

$$\begin{array}{ccc} V & \xrightarrow{(e, r\circ\lambda)} & K(Z, n-1) \times E \\ {\scriptstyle\rho}\downarrow & & \downarrow{\scriptstyle\mu} \\ E & = & E \end{array} .$$

Since $r^*\tilde{\alpha} = 0$, we obtain the information that

$$Sq^2 e = \rho^*\tilde{\alpha} + e \cup \rho^* p^* w_2 .$$

Similarly $Sq^{n-1}e = e \cup \rho^* p^* w_{n-1} + \rho^* x$.

The actual value of x will not matter. We have the following short exact sequence for the Thom complex T ($Z/2$ coefficients are suppressed),

$$0 \to H^{n-1}(E) \xrightarrow{\rho^*} H^{n-1}(V) \xrightarrow{\delta} H^n(T) \to 0$$

with

$$\delta(e) = U.$$

Hence

$$\Theta_n(U) = \Theta_n(\delta e) = [\![\delta(e \cup Sq^2 e)]\!] .$$

Now

$$e \cup Sq^2 e = e \cup (\rho^*\tilde{\alpha} + \rho^* p^*(w_2 w_{n-1})) + \rho^*(x \cdot p^* w_2)$$

and

$$\delta(e \cup \rho^* y) = U \cdot y$$

is a general fact, concluding the proof. □

The papers by Massey [**72**], Thomas [**114**], Mahowald and Peterson [**68**] and Mahowald [**63**] develop this kind of calculation to obtain results about immersions and tangent vector fields.

7.3.3. Twisted secondary operations. We follow the calculation made at the beginning of (7.3.2) with another, leading to the abstract development of operations to be discussed. Consider two maps,

$$\alpha : BSO(n) \ltimes K(Z,n) \to K(Z/2,n+2)$$

with $\alpha^*(b_{n+2}) = w_2 \otimes u_n + 1 \otimes Sq^2 u_n$ and

$$\beta : BSO(u) \ltimes K(Z/2,n+2) \to K(Z/2,n+4)$$

with $\beta^*(b_{n+4}) = w_2 \otimes b_{n+2} + 1 \otimes Sq^2 b_{n+2}$.

Then, using the Cartan formula and the Adem relations, we find that the composition $\beta \circ (1 \ltimes \alpha)$ is null-homotopic;

$$\begin{aligned}(1 \ltimes \alpha)^*(w_2 \otimes b_{n+2} + 1 \otimes Sq^2 b_{n+2}) &= w_2^2 \otimes u_n + w_2 \otimes Sq^2 u_n \\ &\quad + Sq^2(w_2 \otimes u_n + 1 \otimes Sq^2 u_n) \\ &= 0\,.\end{aligned}$$

The algebra may be abstracted in the notion of a semi-tensor algebra structure on

$$H^* BSO(n) \otimes \mathcal{A}$$

and is introduced in [**73**] and [**77**]. Relations in this algebra may be used to produce secondary operations in a manner similar to the use of the Adem relations. Rather than set things up initially with this algebra, we shall develop the subject in a geometric manner, similar to our development of secondary operations in Chapter 4.

Twisted secondary operations are associated with a pair of maps

$$\begin{aligned}\alpha &: D \ltimes C_0 \to C_1\,, \\ \beta &: D \ltimes C_1 \to C_2\end{aligned}$$

with C_2 a simply connected H-space, and a contracting homotopy N for the composition $\beta \circ 1 \ltimes \alpha$. Here

$$1 \ltimes \alpha : D \ltimes C_0 \to D \ltimes C_1$$

is given by $(1 \ltimes \alpha)(x,y) = (x, \alpha(x,y))$ and, as before, the diagonal map on D has been suppressed from the notation.

Given a space X and a fixed map $v : X \to D$ we introduce the source and target for the new operation. We set

$$S_{v,\alpha}(X) = \{[\varepsilon] | \varepsilon : X \to C_0,\ \alpha \circ v \ltimes \varepsilon \sim *\}\,,$$

$$T_{v,\hat{\beta}}(X) = [X, \Omega C_2]/im\, \hat{\beta}_\#$$

where $\hat{\beta}_\# : [X, \Omega C_1] \to [X, \Omega C_2]$ is given by

$$\delta \mapsto \hat{\beta} \circ v \ltimes \delta\,.$$

The semi-additivity formula for $\hat{\beta}$ yields that $\hat{\beta}_{\#}$ is a homomorphism, with respect to loop addition on source and target.

The inclusion of the data $v : X \to D$ is motivated by applications. In the literature, a new category of spaces over D is introduced for the formal management. For the sake of variety, I shall develop the subject in our present category $\mathcal{K}_*$, and carry the extra structure explicitly. One advantage of this approach is a direct link with earlier material.

The naturality properties of source and target are the expected ones for maps over D,

$$\begin{array}{ccc} X' & \xrightarrow{f} & X \\ & {\scriptstyle v'}\searrow \quad \swarrow{\scriptstyle v} & \\ & D\ , & \end{array} \qquad v \circ f \sim v' \ .$$

Thus f induces

$$f^{\#} : S_{v,\alpha}(X) \to S_{v',\alpha}(X')$$

given by

$$f^{\#}(\varepsilon) = \varepsilon \circ f$$

and $f^{\#} : T_{v,\hat{\beta}}(X) \to T_{v',\beta}(X')$ given by $f^{\#}[\![g]\!] = [\![g \circ f]\!]$.

We check that the latter map is well-defined,

$$\begin{aligned} f^{\#}(\hat{\beta}_{\#}(\delta)) &= \hat{\beta} \circ (v \ltimes \delta) \circ f \\ &= \hat{\beta} \circ (v \circ f \ltimes \delta \circ f) \\ &\simeq \hat{\beta} \circ (v' \ltimes \delta \circ f) \\ &= \hat{\beta}_{\#}(f^{\#}(\delta)) \ . \end{aligned}$$

By definition, a twisted secondary operation is a natural transformation

$$\Theta(v, \quad) : S_{v,\alpha}(\qquad) \to T_{v,\hat{\beta}}(\qquad)$$

where naturality is with respect to maps over D. The case $D =$ pt. is just that considered in Chapter 4.

We can base our discussion of twisted operations on the material in Chapter 4 because there is a well-defined map

$$D \ltimes \Omega C \to \Omega(D \ltimes C)$$

induced by the equation

$$(x, \lambda)(t) = (x, \lambda(t)) \ .$$

In particular $\hat{\beta}$ factors through $\Omega\beta$,

$$\begin{array}{ccc} D \ltimes \Omega C_1 & \xrightarrow{\hat{\beta}} & \Omega C_2 \\ \downarrow & \nearrow & \\ \Omega(D \ltimes C_1) & & \Omega\beta \ . \end{array}$$

We have the source for $1 \ltimes \alpha$,

$$S_{1\ltimes\alpha}(X) = \{[\varepsilon] \mid \varepsilon : X \to D \ltimes C_0,\ 1 \ltimes \alpha \circ \varepsilon \sim *\}$$

and

$$S_{v,\alpha}(X) \subset S_{1\ltimes\alpha}(X)$$

as a subset. Moreover, if $v \ltimes \varepsilon$ is in $S_{1\ltimes\alpha}(X)$, then ε is in $S_{v,\alpha}(X)$, by use of the counit. Similarly, the inclusion

$$im\,\hat{\beta}_{\#} \subset im\,\Omega\beta$$

induces a map

$$T_{v,\hat{\beta}}(X) \to T_{\Omega\beta}(X) \ .$$

Here is the basic result of this section. *Given $(\beta, 1 \ltimes \alpha, N)$ a sequence with homotopy, there is a natural transformation*

$$\Theta(v, \quad) : S_{v,\alpha}(X) \to T_{v,\hat{\beta}}(X)$$

such that the following diagram commutes,

$$\begin{array}{ccc} S_{v,\alpha}(X) & \xrightarrow{\Theta(v,\ \ )} & T_{v,\hat{\beta}}(X) \\ \downarrow & & \downarrow \\ S_{1\ltimes\alpha}(X) & \xrightarrow[\Theta]{} & T_{\Omega\beta}(X) \end{array}$$

where Θ is the secondary operation associated with $(\beta, 1 \ltimes \alpha, N)$.

Proof. We construct $\Theta(v, \quad)$ with the diagram

$$\begin{array}{ccccc} X & = & X & = & X \\ \downarrow{\scriptstyle v\ltimes\varepsilon} & & \downarrow & & \downarrow{\scriptstyle v\ltimes H^{\natural}} \\ D \ltimes C_0 & \xrightarrow[1\ltimes\alpha]{} & D \ltimes C_1 & \longleftarrow & D \ltimes PC_1 \\ \downarrow{\scriptstyle N^{\natural}_{\tau}} & & \downarrow{\scriptstyle\beta} & & \downarrow{\scriptstyle\hat{\beta}} \\ P_1C_2 & \longrightarrow & C_2 & \longleftarrow & P_0C_2 \end{array}$$

where H is a contracting homotopy for $\alpha \circ v \ltimes \varepsilon$. Note that v induces a monomorphism

$$[X, C_0] \longrightarrow [X, D \ltimes C_0]$$

$$\varepsilon \longrightarrow v \ltimes \varepsilon$$

because C_0 is a retract of $D \ltimes C_0$. The unnamed vertical map is $v \ltimes (\alpha \circ v \ltimes \varepsilon)$. Now

$$[\![\{\hat{\beta} \circ v \ltimes H^{\natural},\ N^{\natural}_{\tau} \circ v \ltimes \varepsilon\}]\!]$$

is a well-defined element of $T_{v,\hat{\beta}}(X)$ and the result follows. □

The basic properties of secondary operations set out in the fundamental theorem may be passed to twisted operations by our result.

Homotopy invariance is immediate.

The semi-additivity property for $\Theta(v,\quad)$ is a consequence of naturality as the formulation of subsection (4.1.1) works here. We set down the notation. We start with the diagram

$$\begin{array}{ccc} X/Y & \xrightarrow{r} & \Sigma Y \\ {\scriptstyle v}\downarrow & & \downarrow{\scriptstyle v'} \\ D & = & D \end{array} \qquad \text{with } v' \circ r \sim v\ .$$

Given $\varepsilon_1 : X/Y \to C_0$, $\varepsilon_2 : \Sigma Y \to C_0$ such that

$$X/Y \xrightarrow{v \ltimes \varepsilon_1} D \ltimes C_0 \xrightarrow{\alpha} C_1\ ,$$

$$\Sigma Y \xrightarrow{v' \ltimes \varepsilon_2} D \ltimes C_0 \xrightarrow{\alpha} C_1$$

are both null-homotopic, then

$$\varepsilon_1 + \varepsilon_2 \circ r \text{ is in } S_{v,\alpha}(X/Y)$$

with the sum induced by the fold map for C_0. We have a natural homeomorphism

$$D \ltimes (C_0 \vee C_0) \cong D \ltimes C_0 \vee D \ltimes C_0\ .$$

We may write the equation

$$\begin{aligned} \Theta(v, \varepsilon_1 + \varepsilon_2 \circ r) &= \Theta(v,\ \text{fold } \circ (\varepsilon_1 \vee \varepsilon_2) \circ c) \\ &= c^{\#}(\Theta(v, \varepsilon_1),\ \Theta(v', \varepsilon_2)) \\ &= \Theta(v, \varepsilon_1) + \Theta(v, \varepsilon_2 \circ r)\ . \end{aligned}$$

Alternatively, semi-additivity may be inferred from the same property for Θ.

Changes in the contracting homotopy are measured by

$$\delta : D \ltimes C_0 \to \Omega C_2$$

with the resulting twisted secondary operations differing by a twisted primary operation.

Next we develop loop and suspension information for twisted operations. We can work through adjoints or by taking loops and we shall do both, beginning with adjoints. Given

$$\alpha : D \ltimes C_0 \to C_1$$

we have $\hat{\alpha} : D \ltimes \Omega C_0 \to \Omega C_1$.

Now $S_{v,\hat{\alpha}}(X) = \{[\varepsilon] \mid \varepsilon : X \to \Omega C_0 \text{ and } \hat{\alpha} \circ v \ltimes \varepsilon \sim *\}$. Since $\hat{\alpha}$ factors through $\Omega\alpha$, we have the adjoint

$$(v \ltimes \varepsilon)^\flat : \Sigma X \to D \ltimes C_0$$

and $(v \ltimes \varepsilon)^\flat$ is in $S_{1\ltimes\alpha}(\Sigma X)$.

We write $S^*_\alpha(\Sigma X) \subset S_{1\ltimes\alpha}(\Sigma X)$ for the subset of elements obtained this way.

Remark. The functor Σ_D introduced by McClendon can be described in terms of the construction above.

Similarly, we have

$$T^*_{\hat{\beta}}(\Sigma X) = [\Sigma X, \Omega C_2]/(im\,\hat{\beta} \text{ on } (v \ltimes \delta)^\flat)$$

where $\delta : X \to \Omega C_1$, and the natural projection

$$T^*_{\hat{\beta}}(\Sigma X) \to T_{\Omega\beta}(\Sigma X)\,.$$

Moreover, taking adjoints in this way, establishes an equality of sets

$$S_{v,\hat{\alpha}}(X) = S^*_\alpha(\Sigma X),$$

and an isomorphism of abelian groups

$$T_{v,\hat{\beta}}(X) \simeq T^*_{\hat{\beta}}(\Sigma X)\,.$$

Now suppose Θ is an ordinary secondary operation based on $(\beta, 1\ltimes\alpha, N)$. We observe that the image of Θ on $S^*_\alpha(\Sigma X)$ lies in $T^*_{\hat{\beta}}(\Sigma X)$. This fact is apparent from the diagram below,

$$\begin{array}{ccccc}
\Sigma X & = & \Sigma X & = & \Sigma X \\
\Big\downarrow{\scriptstyle (v\ltimes\varepsilon)^\flat} & & \Big\downarrow & & \Big\downarrow{\scriptstyle (v\ltimes\kappa\circ H^\natural)^\flat} \\
D\ltimes C_0 & \xrightarrow[1\ltimes\alpha]{} & D\ltimes C_1 & \longleftarrow & D\ltimes PC_1 \\
\Big\downarrow{\scriptstyle N^\natural_\tau} & & \Big\downarrow{\scriptstyle \beta} & & \Big\downarrow{\scriptstyle \hat\beta} \\
P_1C_2 & \longrightarrow & C_2 & \longleftarrow & P_0C_2
\end{array}$$

where the unmarked vertical arrow is $(v \ltimes (\alpha \circ v \ltimes \varepsilon))^\flat$. The switch of the variables map κ arises in the following way,

$$X \xrightarrow{v\ltimes H^\natural} D \ltimes P\Omega C_1 \xrightarrow{1\ltimes\kappa} D \ltimes \Omega PC_1 \longrightarrow \Omega(D \ltimes PC_1)$$

and take the adjoint.

We have established a commutative diagram,

$$\begin{array}{ccc}
S_{v,\hat\alpha}(X) & \xrightarrow{\Omega\Theta(v,\quad)} & T_{v,\hat{\hat\beta}}(X) \\
\Big\downarrow{\scriptstyle =} & & \Big\downarrow{\scriptstyle \cong} \\
S^*_\alpha(\Sigma X) & \xrightarrow[\Theta]{} & T^*_{\hat\beta}(\Sigma X)
\end{array} ,$$

and, in particular, that $\Omega\Theta(v, \quad)$ is additive.

If $\Theta(v, \quad)$ is associated with $(\beta, 1 \ltimes \alpha, N)$, then $\Omega\Theta(v, \quad)$ is associated with

$$(\hat\beta, 1 \ltimes \tau \circ \hat\alpha, \tau \circ \hat N) .$$

The argument is similar to the ordinary case based on the diagram in (3.2.7). We leave most of the details for the reader. The key substitution is the following. For a contracting homotopy for $\beta \circ 1 \ltimes \alpha$,

$$N : I \times D \ltimes C_0 \to C_2 ,$$

we have

$$\hat N : I \times D \ltimes \Omega C_0 \to \Omega C_2 .$$

Then $(\hat N)^\natural : D \ltimes \Omega C_0 \to P\Omega C_2$ equals

$$\kappa \circ (\hat N^\natural) .$$

We turn to compatibility with exact sequences. We work with maps over D as in

$$\begin{array}{ccccccc} X & \xrightarrow{f} & Y & \xrightarrow{j} & T_f & \xrightarrow{q} & \Sigma X \\ & & \downarrow{\scriptstyle v} & & \downarrow{\scriptstyle v'} & & \\ & & D & = & D & . & \end{array}$$

We have $A : [T_f, C_0] \to [Y, C_1]$ given by

$$\varepsilon \to \alpha \circ (v' \ltimes \varepsilon) \circ j \simeq \alpha \circ (v \ltimes \varepsilon j) .$$

We observe that

$$j^{\#} : \ker A \to S_{1 \ltimes \alpha}(Y)$$

factors through $S_{v,\alpha}(Y)$.

We have $B : [\Sigma Y, D \ltimes C_1] \to [\Sigma X, C_2]$ given by

$$\gamma \to \beta \circ \gamma \circ \Sigma f .$$

Then we have $\Delta = \Delta(\beta, 1 \ltimes \alpha)$

$$\Delta : \ker A \to \ \text{coker}\ B$$

just as in Chapter 4.

The known compatibility of Θ with Δ transfers to $\Theta(v, \quad)$ as in the following commuative diagram

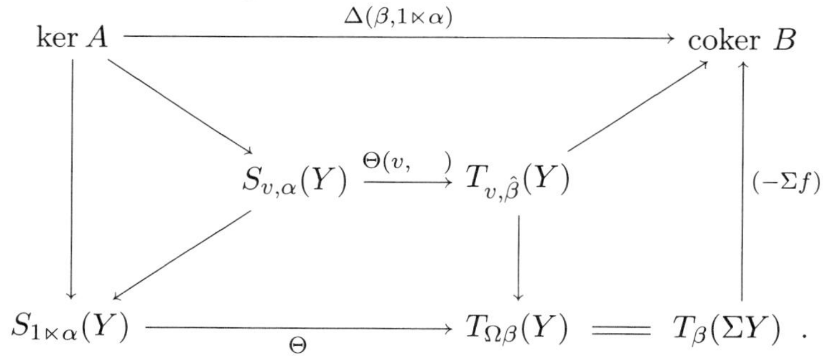

7.3.4. Cartan formulas. Suppose Θ is a secondary operation based on a relation $\varphi\theta = 0$. We discuss the evaluation of Θ on a product in terms of operations on the factors. To see what is involved, we first use a Cartan formula for θ to analyze the equation $\theta(xy) = 0$. We write

$$0 = \sum_i \pm \theta_i' x \cup \theta_i'' y$$

where the sign is $(-1)^{|x|\,|\theta_i''|}$ and we regard θ as an element of the Steenrod algebra, so in particular, Bocksteins are unsigned. Since this equation can hold in a variety of ways, we expect a similar variety in expressions for $\Theta(xy)$. The subject of Cartan formulas for secondary operations has an

<u>ad hoc</u> character and is guided by the facts of particular cases. The general principles are a methodology for constructing Cartan formulas. We follow Thomas [**115**] in its development.

We begin with the generic data for Θ, its universal example, and the universal example for the exterior product $x \wedge y$,

$$\begin{array}{ccc} W_\theta & \xrightarrow{q} & PC_1 \\ {\scriptstyle p}\downarrow & & \downarrow \\ C_0 & \xrightarrow[\theta]{} & C_1 \end{array} , \qquad C_0' \wedge C_0'' \xrightarrow{m} C_0 .$$

The ingredients of the methodology, named <u>Cartan data</u> by Thomas, are the following pair of fiber squares, both pull-backs,

$$\begin{array}{ccc} W' & \xrightarrow{q'} & PD' \\ {\scriptstyle p'}\downarrow & & \downarrow \\ C_0' & \xrightarrow[\alpha']{} & D' \end{array} , \qquad \begin{array}{ccc} W'' & \xrightarrow{q''} & PD'' \\ {\scriptstyle p''}\downarrow & & \downarrow \\ C_0'' & \xrightarrow[\alpha'']{} & D'' \end{array}$$

and a pair of maps

$$\begin{aligned} n' &: D' \wedge C_0'' \to C_1 , \\ n'' &: C_0' \wedge D'' \to C_1 \end{aligned}$$

such that the composition $\theta \circ m$ factors up to homotopy,

$$\begin{array}{ccc} C_0' \wedge C_0'' & \xrightarrow{\quad m \quad} & C_0 \\ {\scriptstyle \Delta}\downarrow & & \downarrow{\scriptstyle \theta} \\ C_0' \wedge C_0'' \times C_0' \wedge C_0'' & & C_1 \\ {\scriptstyle \alpha'\wedge 1 \times 1 \wedge \alpha''}\downarrow & & \uparrow{\scriptstyle g} \\ D' \wedge C_0'' \times C_0' \wedge D'' & \xrightarrow[n' \times n'']{} & C_1 \times C_1 \end{array}$$

and we assume C_1 is an H-space with base point serving as a strict unit for its multiplication, written as g.

The choices for $\alpha', \alpha'', n', n''$ reflect the possible ways for $\theta(xy) = 0$ to hold in the universal example.

In the presence of Cartan data, the composition

$$\theta \circ m \circ p' \wedge p'' : W' \wedge W'' \to C_1$$

is null because it factors through

$$\alpha' p' \wedge p'' \times p' \wedge \alpha'' p'' .$$

Hence, there is a map

$$\ell : W' \wedge W'' \to W_\theta$$

such that $p \circ \ell \cong m \circ p' \wedge p''$. A Cartan formula for Θ is encoded in the composition $\tilde{\varphi} \circ \ell$.

To analyze ℓ, let H be a homotopy from

$$g(n' \circ \alpha' \wedge 1,\ n'' \circ 1 \wedge \alpha'') \circ \Delta$$

to

$$\theta \circ m\,.$$

Write elements of W' as (x', λ') such that $\lambda'(1) = \alpha'(x')$ and likewise for W''. Then we have

$$\ell(x', \lambda', x'', \lambda'') = (m(x', x''), \{\hat{g}(\hat{n}'(\lambda', x''), \hat{n}''(x', \lambda'')) + H^\natural(x', x'')\})$$

Here

$$\hat{g} : PC_1 \times PC_1 \to PC_1$$

is induced by g,

$$\hat{g}(\lambda_1, \lambda_2)(s) = g(\lambda_1(s), \lambda_2(s))\,,$$

and

$$\begin{aligned} \hat{n}' &: PD' \wedge C_0'' \to PC_1\,, \\ \hat{n}'' &: C_0' \wedge PD'' \to PC_1 \end{aligned}$$

are the semi-evaluation maps.

With this construction, we have

(i) "Stable range Cartan formula." The following diagram commutes up to homotopy,

$$\begin{array}{ccccc} W' \wedge \Omega D'' & \xrightarrow{1\wedge j''} & W' \wedge W'' & \xleftarrow{j'\wedge 1} & \Omega D' \wedge W'' \\ & & \downarrow{\scriptstyle \ell} & & \\ \downarrow{\scriptstyle p'\wedge 1} & & W_\theta & & \downarrow{\scriptstyle 1\wedge p''} \\ & & \uparrow{\scriptstyle j} & & \\ C_0' \wedge \Omega D'' & \xrightarrow[\hat{n}'']{} & \Omega C_1 & \xleftarrow[\hat{n}']{} & \Omega D' \wedge C_0''\,. \end{array}$$

Proof. The verification is direct;

$$\begin{aligned} \ell \circ 1 \wedge j'(x', \lambda', \lambda_2) &= \ell(x', \lambda', *, \lambda_2) \\ &= (m(x', *), \{\hat{g}(\hat{n}'(\lambda', *), \hat{n}''(x', \lambda_2) + H^\natural(x', *)\})\,. \end{aligned}$$

Now $\hat{n}'(\lambda', *) = *$, which is a strict identity for C_1. We have

$$\{\hat{g}(*, \hat{n}''(x', \lambda_2)) + *\}(s) = \begin{Bmatrix} n''(x', \lambda_2(2s)) & 0 \le s \le \frac{1}{2} \\ * & \frac{1}{2} \le s \le 1 \end{Bmatrix} .$$

Thus the homotopy commutativity for the left square follows. The right square is handled similarly. We note in passing that

$$\ell \circ j' \wedge j'' \sim *$$

which limits the form of middle terms in the "stable range" and in the presence of Cartan data. □

To illustrate the formalism, we consider the Adams operation $\varphi_{1,1}$ based on

$$Sq^2 Sq^2 + Sq^3 Sq^1 = 0 .$$

We write $K(m)$ for $K(Z/2, m)$. The universal examples are

$$\begin{array}{ccc} W_{m+n} & \xrightarrow{\tilde{\varphi}} & K(m+n+3) \\ \downarrow & & \\ K(m+n) & \xrightarrow[\substack{Sq^1 \\ Sq^2}]{} & K(m+n+1, m+n+2) \end{array}$$

and $K_m \wedge K_n \to K_{m+n}$ represents $b_m \otimes b_n$.

We shall develop a formula for $\varphi_{1,1}(xy)$ when each factor is in the domain of $\varphi_{1,1}$.

We write

$$\alpha' : K(m) \to K(m+1, m+2) \text{ for } \begin{pmatrix} Sq^1 \\ Sq^2 \end{pmatrix}$$

and similarly for α''. We write e', e'' for representatives of $\varphi_{1,1}$, e.g.

$$e' : W_m \to K(m+3) .$$

Let $n' : K(m+1, m+2) \wedge K_n \to K(m+n+1, m+n+2)$ represent

$$\begin{pmatrix} b_{m+1} \otimes 1 \otimes b_n \\ 1 \otimes b_{m+2} \otimes b_n \end{pmatrix} ,$$

and let

$$n'' : K(m) \wedge K(n+1, n+2) \to K(m+n+1, m+n+2)$$

represent

$$\begin{pmatrix} b_m \otimes b_{n+1} \otimes 1 \\ Sq^1 b_m \otimes b_{n+1} \otimes 1 + b_m \otimes 1 \otimes b_{n+2} \end{pmatrix} .$$

Now $\theta \circ m$ represents

$$\begin{pmatrix} Sq^1 b_m \otimes b_n + b_m \otimes Sq^1 b_n \\ Sq^2 b_m \otimes b_n + Sq^1 b_m \otimes Sq^1 b_n + b_m \otimes Sq^2 b_n \end{pmatrix} .$$

Moreover, $n' \circ \alpha' \wedge 1$ represents

$$\begin{pmatrix} Sq^1 b_m \otimes b_n \\ Sq^2 b_m \otimes b_n \end{pmatrix}$$

and $n'' \circ 1 \wedge \alpha''$ represents

$$\begin{pmatrix} b_m \otimes Sq^1 b_n \\ Sq^1 b_m \otimes Sq^1 b_n + b_m \otimes Sq^2 b_n \end{pmatrix} .$$

Thus their sum is $\theta \circ m$. Next we calculate

$$\tilde{\varphi} \circ j_{m+n} \circ \hat{n}'' \circ (p_m \wedge 1).$$

We obtain

$$\begin{aligned}(p_m \wedge 1)^*(Sq^3(b_m \otimes b_n \otimes 1) + Sq^2(Sq^1 b_m \otimes b_n \otimes 1 + b_m \otimes 1 \otimes b_{n+1})) \\ = p_m^* b_m \otimes (Sq^3 b_n + Sq^2 b_{n+1}),\end{aligned}$$

because $p_m^*(\gamma b_m) = 0$ if $|\gamma| > 0$.

Thus $\tilde{\varphi} \circ \ell$ has $p_m^* b_m \otimes e''$ as a summand. Working the other side of the square in (i) gives

$$e' \otimes p_n^* b_n .$$

Now, from the cohomology structure of the universal examples, we see that each of

$$j_{m+n}^*,\ j_m^*,\ j_n^*$$

are monomorphisms respectively in dimensions

$$m+n+3,\ m+3,\ n+3 ,$$

provided $m, n > 2$. Thus $\tilde{\varphi} \circ \ell$ represents

$$p_m^* b_m \otimes e'' + e' \otimes p_n^* b_n .$$

In the usual notation, we have

$$\varphi_{1,1}(xy) = \varphi_{11}(x) \cdot y + x \cdot \varphi_{11}(y) \text{ in } H^{m+n+3}(X)/Ind(\varphi_{1,1}) .$$

The example illustrates a general fact for stable operations based on an Adem relation $\varphi\theta = 0$, and where each factor is annihilated by Steenrod operations of positive degree [**6**];

$$\Theta(xy) = \Theta(x) \cdot y + (-1)^{d|x|} x \cdot \Theta(y) \text{ in } H^*(X)/Ind(X, \Theta)$$

where $d = |\varphi| + |\theta| - 1 = \deg(\Theta)$. We verify the sign. The secondary operation Θ has components Θ_m based on $(\varphi, \tau_m \circ \theta)$ where φ, θ are in $\mathcal{A}$. We have

$$\theta(b_m \otimes b_n) = \theta b_m \otimes b_n + \cdots + (-1)^{m|\theta|} b_m \otimes \theta b_n \ .$$

We may construct

$$\begin{array}{ccc} K(m) \wedge K(n) & \longrightarrow & K(m+n) \\ {\scriptstyle 1\wedge\alpha''}\downarrow & & \downarrow{\scriptstyle \tau_{m+n}\circ\theta} \\ K(m) \wedge K(n+|\theta|) & \underset{n''}{\longrightarrow} & K(m+n+|\theta|) \end{array}$$

with α'' representing $\tau_n \circ \theta$ and n'' representing

$$(-1)^{m|\theta|} \tau_m b_m \otimes b_{n+|\theta|} + \text{ other terms } \ .$$

Now $\Omega\varphi$ on the displayed element gives

$$(-1)^{m|\theta|} \tau_m (-1)^{m|\varphi|} b_m \otimes \Omega\varphi(b_{n+|\theta|-1}) \ ,$$

and substituting $\tau_m = (-1)^m$ gives the stated value of d. $\square$

To continue the general discussion of Cartan formulas, we introduce a refinement of the Cartan data which will permit a direct analysis of $\tilde{\varphi} \circ \ell$, without first composing with $j' \wedge 1$ or $1 \wedge j''$. We shall obtain complete information on the summands in $\tilde{\varphi} \circ \ell$ modulo terms from $C_0' \wedge C_0''$ without imposing dimensional restrictions.

We have discussed certain maps n', n'' as part of a factorization of $\theta \circ m$. We now look at $\varphi n', \varphi n''$. We introduce a capitalization of Cartan data, consisting of

$$\begin{aligned} \beta' &: D' \to E' \text{ such that } \beta'\alpha' \sim * \ , \\ \beta'' &: D'' \to E'' \text{ such that } \beta''\alpha'' \sim * \ , \\ N' &: E' \wedge C_0'' \to C_2 \text{ such that } \varphi n' \sim N' \circ \beta' \wedge 1 \ , \\ N'' &: C_0' \wedge E'' \to C_2 \text{ such that } \varphi n'' \sim N'' \circ 1 \wedge \beta'' \ . \end{aligned}$$

Suppose C_2 is an H-space, with basepoint serving as a strict unit

$$h : C_2 \times C_2 \to C_2$$

and φ is an H-map. Then the following diagram is homotopy commutative up to a map

$$\mathcal{E} \circ p' \wedge p'' \ , \quad \mathcal{E} : C_0' \wedge C_0'' \to \Omega C_2,$$

(ii)

$$\begin{array}{ccccc}
W' \wedge W'' & \xrightarrow{\ell} & W_\theta & \xrightarrow{\tilde{\varphi}} & \Omega C_2 \\
\Big\downarrow{\scriptstyle\Delta} & & & & \Big\uparrow{\scriptstyle\hat{h}} \\
W' \wedge W'' \times W' \wedge W'' & & & & \\
\Big\downarrow{\scriptstyle\tilde{\beta}'\wedge p'' \times p' \wedge \tilde{\beta}''} & & & & \\
\Omega E' \wedge C_0'' \times C_0' \wedge \Omega E'' & & \xrightarrow[\hat{N}' \times \hat{N}'']{} & & \Omega C_2 \times \Omega C_2 \ .
\end{array}$$

Here $\tilde{\beta}'$ and $\tilde{\beta}''$ are coliftings based on contracting homotopies for $\beta'\alpha'$ and $\beta''\alpha''$,

$$(\beta', \alpha', L') \ , \quad (\beta'', \alpha'', L'') \ ,$$

together with (φ, θ, L), and

$$\hat{h}(\lambda_1, \lambda_2)(s) = h(\lambda, (s), \lambda(s)) \ .$$

We can say more about the map $\mathcal{E}$. It depends on all the homotopies involved. Thus, in addition to the contracting homotopies L', L'', we write

$$M', \text{ a homotopy from } N' \circ \beta' \wedge 1 \text{ to } \varphi n'$$

and

$$M'', \text{ a homotopy from } N'' \circ 1 \wedge \beta'' \text{ to } \varphi n'' \ ,$$

and

$$K, \text{ a homotopy from } h \circ \varphi \times \varphi \text{ to } \varphi g \ .$$

We have

$$\hat{M}', \text{ a homotopy from } \hat{N}' \circ P\beta' \wedge 1 \text{ to } P\varphi \circ \hat{n}'$$

given by

$$\hat{M}'(s, \lambda', c'')(t) = M'(s, \lambda'(t), c'') \ ,$$

and similarly for $\hat{M}''$ from $\hat{N}'' \circ 1 \wedge P\beta''$ to $P\varphi \circ \hat{n}''$. Likewise, we have $\hat{K}$ from $\hat{h} \circ P\varphi \times P\varphi$ to $P\varphi \circ \hat{g}$ given by

$$\hat{K}(s, \lambda_1, \lambda_2)(t) = K(s, \lambda_1(t), \lambda_2(t)) \ .$$

Finally, write $A = (\alpha' \wedge 1 \times 1 \wedge \alpha'') \circ \Delta$. In terms of these maps and homotopies, we will establish that

$$\mathcal{E} = \{\hat{h}(\hat{N}' \circ L' \wedge 1,\ \hat{N}'' \circ 1 \wedge L''),\ \hat{h}(\hat{M}', \hat{M}'') \circ A,\ K^\natural(n', n'') \circ A, P\varphi H^\natural, L_\tau^\natural \circ m\}$$

as a string of 5 homotopies.

With this $\mathcal{E}$, we establish (ii) by direct verification. Let

$$S : W' \wedge W'' \to PD' \wedge C_0'' \times C_0' \wedge PD''$$

be the composite

$$(q' \wedge p'' \times p' \wedge q'') \circ \Delta$$

where the p's and q's are from the fiber squares for the W's. The following picture displays the homotopies mediating between $\mathcal{E}$ and the two ways around the diagram in (ii),

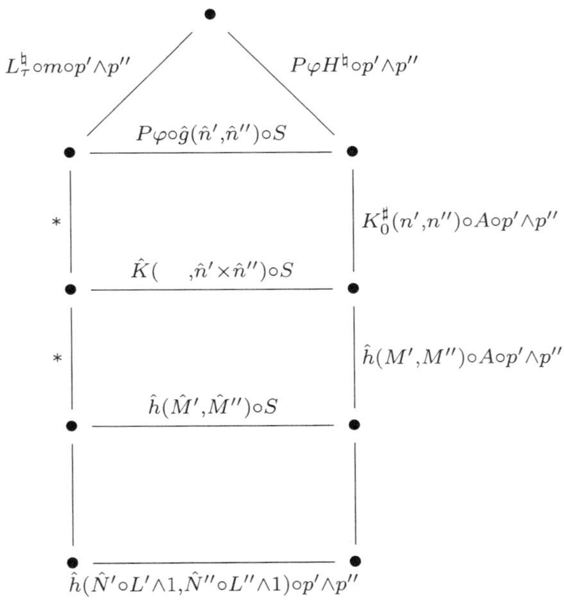

To understand this picture, the "attic" is $\tilde{\varphi} \circ \ell$ expressed as a loop. The "basement" is the other way around the diagram in (ii). The first and second "floors" are filled in with the indicated homotopies. The loop $\mathcal{E}$ is obtained by traversing the boundary of this "house".

We write this out in terms of strings of homotopies. So

$$\tilde{\varphi} \circ \ell = \{P\varphi \circ \hat{g} \circ (\hat{n}' \times \hat{n}'') \circ S,\ P\varphi \circ H^\natural \circ p' \wedge p'', L_T^\natural \circ m \circ p' \wedge p''\}.$$

The homotopy $\hat{K}(\quad, \hat{n}' \times \hat{n}'') \circ S$ connects $\tilde{\varphi} \circ \ell$ with

$$\{\hat{h}(P\varphi \circ \hat{n}',\ P\varphi \circ \hat{n}'') \circ S,\ K^\natural(n', n'') \circ A \circ p' \wedge p'',\\ P\varphi H^\natural \circ p' \wedge p'',\ L_T^\natural \circ m \circ p' \wedge p''\}\ .$$

The homotopy $\hat{h}(\hat{M}', \hat{M}'') \circ S$ connects this string with

$$\Big\{\hat{h}(\hat{N}' \circ P\beta' \wedge 1,\ \hat{N}'' \circ 1 \wedge P\beta'') \circ S,\ \hat{h}(M', M'') \circ A \circ p' \wedge p'',\\ K^\natural(n', n'') \circ A \circ p' \wedge p'',\ P\varphi \circ H^\natural \circ p' \wedge p'',\ L_T^\natural \circ m \circ p' \wedge p''\Big\}\ .$$

Thus $\tilde{\varphi} \circ \ell$ is homotopic to the last displayed map

$$W' \wedge W'' \to \Omega C_2,$$

and is a traverse of the outside of the "house" neglecting the "basement." Next we look in the "basement." Now

$$\begin{aligned}
&(\tilde{\beta}' \wedge p'' \times p' \wedge \tilde{\beta}'') \circ \Delta \\
&= (P\beta' q' + L'_\tau \circ p') \wedge p'' + p' \wedge (P\beta'' \circ q'' + L''_\tau \circ p'') \\
&= (P\beta' \wedge 1 \times 1 \wedge P\beta'') \circ S + (L'_\tau \wedge 1 \times 1 \wedge L''_\tau) \circ p' \wedge p'' \,.
\end{aligned}$$

Using the semi-additivity of $\hat{N}', \hat{N}''$, we can resolve the "basement" loop as

$$\hat{h} \circ \hat{N}' \times \hat{N}'' \circ (\tilde{\beta}' \wedge p'' \times p' \wedge \tilde{\beta}'') \circ \Delta + \mathcal{E} \circ p' \wedge p''$$

and this completes the proof of (ii). □

To illustrate the methodology of (ii), we work out a Cartan formula for the operation studied in the first section of Chapter 5. Kristensen has treated this operation in [**55**], [**56**]. His method gives more precise results. There is unpublished, prior work of Mahowald, Peterson, and Richter on this operation and I am grateful to have received a copy of their manuscript.

We shall write $\Theta_{n,k}$ for an operation based on the null composition

$$K(Z, n+k) \xrightarrow{Sq^n} K(Z/2, 2n+k) \xrightarrow{Sq^2} K(Z/2, 2n+k+2)$$

and restrict k to 0 or 1 so as not to restrict n. To begin with, we shall not impose restrictions on the tetherings. Suppose x_m, y_n are mod 2 reductions of integral classes with dimensions m, n respectively and

$$y_n^2 = 0 \,.$$

Then we shall establish that

(iii)

$$\begin{aligned}
\Theta_{m+n,0}(xy) &= Sq^m x \cup \Theta_{n,0}(y) + Sq^{m-1} x \cup \Omega\Theta_{n+1,0}(y) \\
&\quad + \Omega\Theta_{m+1,0}(x) \cup Sq^{n-1} y \\
&\quad + \text{ terms from } PH^*K(Z,m) \otimes PH^*K(Z,n) \\
&\quad \text{in } H^{2m+2n+1}(X)/Ind(X, \Theta_{m+n}) \,.
\end{aligned}$$

We first show that

$$\Theta_{m+n,0}(xy) = Sq^m x \cup \Theta_{n,0}(y) + \text{ terms from } H^{2m+2n+1}(K(Z,m) \wedge K(Z,n))$$
modulo $Ind(X, \Theta)$.

For this, the following data suffices:

$$C_0' = K(Z,m),\ \alpha' = \text{ const. map },$$
$$C_0'' = K(Z,n),\ D'' = K(Z/2,2n),\ \alpha'' = Sq^n ,$$
$$C_0 = K(Z,m+n),\ m : C_0' \wedge C_0'' \to C_0 \text{ represents } b_m \otimes b_n ,$$
$$n'' : C_0' \wedge D'' \to C_1 = K(Z/2, 2m+2n) \text{ represents } Sq^m u_m \otimes b_{2n} .$$

Then the equation to be verified is $\theta \circ m \simeq n'' \circ 1 \wedge \alpha''$, and this holds because

$$Sq^{m+n}(b_m \otimes b_n) = Sq^m u_m \otimes Sq^n u_n .$$

Thus we have Cartan data. We can then capitalize with

$N'' : C_0' \wedge K(Z/2, 2n+2) \to K(Z/2, 2m+2n+2)$ representing $Sq^m u_m \otimes b_{2n+2}$.

Then to check $\varphi n'' \simeq N'' \circ (1 \wedge \beta'')$, we have

$$Sq^2(Sq^m u_m \otimes b_{2n}) = Sq^m u_m \otimes Sq^2 b_{2n}$$

and $Sq^2 b_{2n}$ is β''. We can now apply (ii) to give the formula.

To obtain some information about the terms from $K(Z,m) \wedge K(Z,n)$, we consider related operations with the dimensions shifted by one. This step is suggested because the zero property was used in the verification of capitalization. First we look at $\Theta_{m+n,1}$ on $x_m \cdot y_{n+1}$. Since the Cartan formula gives

$$Sq^{m+n}(x \cdot y) = Sq^m x \cdot Sq^n y + Sq^{m-1} x \cdot Sq^{n+1} y$$

we assume both $Sq^n y = 0 = Sq^{n+1} y$. These are independent conditions if and only if n is odd. We continue to work with integral classes or their mod 2 reductions.

Suitable data is on display below, for n odd,

$$\begin{array}{ccc}
K(Z,m) \wedge K(Z,n+1) & \xrightarrow{b_m \otimes b_{n+1}} & K(Z,m+n+1) \\
\downarrow{\scriptstyle 1\wedge\alpha''} & & \downarrow{\scriptstyle Sq^{m+n}} \\
K(Z,m) \wedge K(2n+1,2n+2) & \xrightarrow{n''} & K(2m+2n+1) \\
\downarrow{\scriptstyle 1\wedge\beta''} & & \downarrow{\scriptstyle Sq^2} \\
K(Z,m) \wedge K(2n+3,2n+4) & \xrightarrow[N'']{} & K(2m+2n+3)
\end{array}$$

with

$$\alpha'' = \begin{pmatrix} Sq^n \\ Sq^{n+1} \end{pmatrix} , \quad \beta'' = \begin{pmatrix} Sq^2, & \epsilon Sq^1 \\ 0 & Sq^2 \end{pmatrix}$$

with $\epsilon = 1$ if m is odd and 0 otherwise, and n'' represents

$$Sq^m u_m \otimes b_{2n+1} + Sq^{m-1} u_m \otimes b_{2n+2} ;$$

for n even, we replace α'', β'' with

$$K(Z,n+1) \xrightarrow[Sq^n]{} K(2n+1) \xrightarrow[\binom{Sq^2}{Sq^2Sq^1}]{} K(2n+3,2n+4)$$

and n'' represents

$$Sq^m u_m \otimes b_{2n+1} + Sq^{m-1} u_m \otimes Sq^1 b_{2n+1} \ .$$

For n odd, N'' represents

$$Sq^m u_m \otimes b_{2n+3} + Sq^{m-1} u_m \otimes b_{2n+4} \ ,$$

and for n even,

$$Sq^m u_m \otimes b_{2n+3} + Sq^{m-1} u_m \otimes b_{2n+4} \ .$$

It is routine to check homotopy commutativity.

We can now apply (ii) to obtain the equation, for n odd,

$$\begin{aligned}\Theta_{m+n,1}(x_m \cdot y_{n+1}) = Sq^m x_m \cup \Phi_{n,1}(y_{n+1}) + Sq^{m-1} x_m \cup \Theta_{n+1,0}(y_{n+1}) \\ + \text{ terms from } H^{2m+2n+2}(K(Z,m) \wedge K(Z,n+1))\end{aligned}$$

where $\Phi_{n,1}$ is based on $(Sq^2, Sq^1)\binom{Sq^n}{Sq^{n+1}} = 0$ and we have tacitly assumed m is odd, the more complex case. In the case where $y_{n+1} = \Delta^* y_n$, we have

$$\Delta^* \Theta_{m+n,0}(x_m \cdot y_n) = \Theta_{m+n,1}(x_m \cdot \Delta^* y_n),$$

by using the map

$$K_m \wedge \Sigma K_n \to K_m \wedge K_{n+1} \to K_{m+n+1} \ .$$

Now $\Theta_{n+1,0}(\Delta^* y_n) = \Delta^* \Omega \Theta_{n+1,0}(y_n)$ and we can write

$$\begin{aligned}\Theta_{m+n,0}(xy) = Sq^m x \cup \Omega\Phi_{n,1}(y) + Sq^{m-1} x \cup \Omega\Theta_{n+1,0}(y) \\ + \text{ terms from } \bar{H}^* K(Z,m) \otimes PH^* K(Z,n) \ .\end{aligned}$$

Next, we use the coproduct theorem to compare $\Theta_{n,0}$ with $\Omega\Phi_{n,1}$. The universal example for $\Phi_{n,1}$ is displayed below,

$$\begin{array}{c} E \\ \downarrow \\ K(Z,n+1) \xrightarrow[\binom{Sq^n}{Sq^{n+1}}]{} K(2n+1,2n+2) \ . \end{array}$$

Thus $\Omega E \cong W \times K(Z/2, 2n)$ where W is the universal example for $\Theta_{n,0}$ and the fundamental class b_{2n} has coproduct

$$1 \otimes b_{2n} + p^* u_n \otimes p^* u_n + b_{2n} \otimes 1 \ .$$

Hence $Sq^1 b_{2n}$ is primitive, and we obtain

$$\Omega\Phi_{n,1}(y) = \Theta_{n,0}(y) \text{ modulo } Ind(\Omega\Phi_{n,1}) .$$

Working modulo $Ind(\Theta_{n,0})$, we have

$$\Omega\Phi_{n,1}(y) = \Theta_{n,0}(y) + Sq^1 H^{2n} .$$

Now, m is odd, so we have

$$Sq^m x_m \cup Sq^1 z = Sq^2(Sq^{m-1}x \cdot z) + Sq^{m-1}x \cdot Sq^2 z .$$

Hence the difference between $\Omega\Phi_{n,1}(y)$ and $\Theta_{n,0}(y)$ disappears modulo $Ind(\Theta_{m+n,0})$.

Turning to the case for n even, our formula for N'' indicates we must compare $\Theta_{n+1,0}$ with an operation based on

$$(Sq^2 Sq^1)(Sq^n) .$$

The homotopy commutative diagram

$$\begin{array}{ccc}
K(Z, n+1) & = & K(Z, n+1) \\
\downarrow{\scriptstyle Sq^n} & & \downarrow{\scriptstyle Sq^{n+1}} \\
K(2n+1) & \xrightarrow{Sq^1} & K(2n+2) \\
\downarrow{\scriptstyle Sq^2 Sq^1} & & \downarrow{\scriptstyle Sq^2} \\
K(2n+4) & = & K(2n+4)
\end{array}$$

reveals that these operations agree, using the principles of (4.2.5c).

We now use $\Theta_{m+n,1}$ on $x_{n+1} \cdot y_n$ to obtain the remaining information asserted in (iii). Since

$$Sq^{m+n}(x_{m+1}y_n) = Sq^{m+1}x_{m+1} \cdot Sq^{n-1}y_n + Sq^m x_{m+1} \cdot Sq^n y_n ,$$

we can work with the requirements that

$$Sq^{m+1}x_{m+1} = 0 = Sq^n y_n .$$

Then suitable data for a Cartan formula for $\Theta_{m+n,1}(x_{m+1} \cdot y_n)$ is on display below,

$$\begin{array}{ccc} K(Z,m+1) \wedge K(Z,n) & \xrightarrow{b_{m+1} \otimes b_n} & K(Z,m+n+1) \\ \downarrow{\scriptstyle Sq^{m+1} \wedge 1 \times 1 \wedge Sq^n} & & \downarrow{\scriptstyle Sq^{m+n}} \\ K(2m+2) \wedge K(Z,n) \times K(Z,m+1) \wedge K(2n) & \xrightarrow{n' \times n''} & K(2m+2n+1) \\ \downarrow{\scriptstyle \beta' \wedge 1 \times 1 \wedge \beta''} & & \downarrow{\scriptstyle Sq^2} \\ A & \xrightarrow{N' \times N''} & K(2m+2n+3) \end{array}$$

where A is the space $K(2m+3, 2m+4) \wedge K(Z,n) \times K(Z,m+1) \wedge K(2n+1, 2n+2)$, with n' representing $b_{2m+2} \otimes Sq^{n-1}u_n$,

$$n'' \text{ representing } Sq^m u_{m+1} \otimes b_{2n} \ ,$$

both β' and β'' are $\begin{pmatrix} Sq^1 \\ Sq^2 \end{pmatrix}$ and N' represents $b_{2m+3} \otimes Sq^1 Sq^{n-1} u_n + b_{2m+4} \otimes Sq^{n-1} u_n$,

$$N'' \text{ represents } Sq^1 Sq^m u_{m+1} \otimes b_{2n+1} + Sq^m u_{m+1} \otimes b_{2n+2} \ .$$

We do not incorporate the Adem relation in order to avoid breaking into cases for m, n. The result from (iii) reads

$$\begin{aligned} \Theta_{m+n,1}(x_{m+1} y_n) &= \Theta_{m+1,0}(x_{m+1}) \cup Sq^{n-1} y_n \\ &\quad + \Psi_{m+1}(x_{m+1}) \cup Sq^1 Sq^{n-1} y_n + Sq^1 Sq^m x_{m+1} \cup \Psi_n(y_n) \\ &\quad + Sq^m u_{m+1} \cup \Theta_{n,0}(y_n), \end{aligned}$$

where Ψ_k is based on $Sq^1 Sq^k = 0$ on classes of dim k, and the equation is modulo terms from

$$H^{2m+2n+2}(K(Z,m+1) \wedge K(Z,n)) \ .$$

As before, substitution of $\Delta^* x_m \cdot y_n$ picks up

$$\Omega\Theta_{m+1,0}(x_m) \cup Sq^{n-1} y_n$$

modulo terms from

$$PH^*(K(Z,m)) \otimes \bar{H}^*(K(Z,n)) \ .$$

Moreover, the summands involving Ψ also involve primary operations which give 0 under our original assumptions on x_m, y_n. Thus we have (iii).

A, perhaps peculiar, feature of this calculation is that the two uses of $\Theta_{m+n,1}$ reveal one of the summands for (iii) while hiding the other. That's the way things are for this approach. The working of Kristensen [**66**] based on cochains gives greater precision, should that be necessary.

Finally, we may invoke Prop. 5.1.3 to substitute

$$y \cup Sq^2 y \text{ for } \Omega\Theta_{n+1,0}(y)$$

and

$$x \cup Sq^2 x \text{ for } \Omega\Theta_{m+1,0}(x)$$

in the statement of (iii). This implies some restrictions on the tetherings, so there are implied limitations on possible changes in the term $\mathcal{E}$ in (ii). On the other hand, in order for the non-explicit terms to have any impact in an application, there would have to be further non-trivial action by Steenrod operations. In practice, one would choose to exhaust that information before turning to the tool at hand here. On that note, we conclude this chapter.

Bibliography

[1] J. F. Adams, *On the structure and applications of the Steenrod algebra*, Comment. Math. Helv. **32** (1958), 180–214.

[2] ———, *On the non-existence of elements of Hopf invariant one*, Ann. of Math. **72** (1960), 20–104.

[3] ———, *Stable Homotopy Theory*, Lecture Notes in Mathematics, Springer-Verlag **3** (1964).

[4] J. Adem, *The relations on Steenrod powers of cohomology classes*, In: *Algebraic Geometry and Topology. A Symposium in Honor of S. Lefschetz.* R. H. Fox, D. C. Spencer, A. W. Tucker (eds), 191–238, Princeton Mathematical Series, **12**, Princeton Univ. Press (1957).

[5] ———, *Un criterio cohomologico para determinar composciones esenciales de transformaciones*, Bol. Soc. Mat. Mexicana **1** (1956), 38–48.

[6] ———, *Sobre operaciones cohomologicos secundarias*, Bol. Sci. Mat. Mexicana **7** (1962), 95–110.

[7] A. Adem, R. J. Milgram, *Cohomology of Finite Groups*, Grundlehren der Mathematischen Wissenschaften, **309**, Springer-Verlag, Berlin (1994).

[8] W. D. Barcus, *On a theorem of Massey and Peterson*, Quart. Jr. Math. **19** (1968), 33–41.

[9] M. G. Barratt, I. M. James, N. Stein, *Whitehead products and projective spaces.* J. Math. Mech. **9** (1960), 813–819.

[10] M. G. Barratt, J. D. S. Jones, M. E. Mahowald, *The Kervaire invariant problem*, Contemporary Mathematics, **19** (1983), 9–22.

[11] M. G. Barratt, M. E. Mahowald, M. C. Tangora, *Some differentials in the Adams spectral sequence - II*, Topology **9** (1970), 309–316.

[12] I. Berstein, J. R. Harper, *Cogroups which are not suspensions*, In: *Algebraic Topology*, G. Carlsson, R. L. Cohen, H. R. Miller, D. C. Ravenel (eds), 63–86, Lecture Notes in Mathematics, Springer-Verlag **1370** (1989).

[13] W. Browder, *Torsion in H-spaces*, Ann. of Math. **74** (1961), 24–51.

[14] ———, *On differential Hopf algebras*, Trans. Amer. Math. Soc. **109** (1963), 153–178.

[15] ———, *The Kervaire invariant of framed manifolds and its generalization*, Ann. of Math. **90** (1969), 157–186.

[16] W. Browder, E. Thomas, *On the projective plane of an H-space*, Illinois Jr. Math. **7** (1963), 492–502.

[17] E. H. Brown, F. P. Peterson, *Whitehead products and cohomology operations*, Quart. Jr. Math. **15** (1964), 116–120.

[18] ———, *The Kervaire invariant of $(8k+2)$-manifolds*, Amer. Jr. Math. **88** (1966), 815–826.

[19] G. Carlsson, R. J. Milgram, *Stable homotopy and iterated loop spaces*, In: *Handbook of Algebraic Topology*, I.M. James, ed., 505–584, Elsevier, Amsterdam (1995).

[20] H. Cartan, *Algebres d'Eilenberg-Maclane et homotopie, Séminaire Henri Cartan 1954–55*, Secrétariat mathématique, Paris (1955).

[21] ———, *Invariant de Hopf et opérations cohomologigies secondaires, Seminaire Henri Cartan 1958–59*, Secrétariat mathématique, Paris (1959).

[22] ———, *Sur l'itération des opérations de Steenrod*, Comment. Math. Helv. **29** (1955), 40–58.

[23] H. Cartan, S. Eilenberg, *Homological Algebra*, Princeton University Press (1956).

[24] F. R. Cohen, *Splitting certain suspensions via self-maps*, Illinois Jr. Math. **20** (1976), 336–347.

[25] ———, *A course in some aspects of classical homotopy theory* In: *Algebraic Topology*, H.R. Miller, D.C. Ravenel (eds), 1–92, Lecture Notes in Mathematics **1286**, Springer-Verlag (1987).

[26] F. R. Cohen, T. J. Lada, J. P. May, *The homology of iterated loop spaces*, Lecture Notes in Mathematics **533**, Springer-Verlag (1976).

[27] B. G. Cooper, *Coproducts in the mod p cohomology of stable two-stage Postnikov systems*, Ph.D. dissertation, Yale University, New Haven, Conn. (1971).

[28] A. Dold, R. Lashof, *Principal quasi-fibrations and fiber homotopy equivalence of bundles*, Illinois Jr. Math. **3** (1959), 285–305.

[29] T. Ganea, *A generalization of the homology and homotopy suspension*, Comment. Math. Helv. **39** (1965), 295–322.

[30] ———, *Cogroups and suspensions*, Invent. Math. **9** (1970), 185–197.

[31] S. Gitler, *Operaciones cohomológicas de order superior*, In: *Lectures on Algebraic and Differential Topology*, R. Bott, S. Gitler, I.M. James (eds). Lecture Notes in Mathematics **279** Springer-Verlag (1972).

[32] S. Gitler, M. E. Mahowald, R. J. Milgram, *Secondary cohomology operations and complex vector bundles*, Proc. Amer. Math. Soc. **22** (1969), 223–229.

[33] S. Gitler, R. J. Milgram, *Evaluating secondary operations on low dimensional classes*, In: *Conference on Algebraic Topology*, 47–60, University of Illinois at Chicago Circle (1968).

[34] S. Gitler, J. D. Stasheff, *The first exotic class of BF*, Topology **4** (1965), 257–266.

[35] P. Goerss, M. E. Mahowald, *Immersions not regularly homotopic to embeddings*, Amer. Jr. Math. **109** (1987), 1171–1195.

[36] D. C. Goncalves, *Mod 2 homotopy associative H-spaces* In: *Geometric Applications of Homotopy Theory 1*, 198–216, Edited by M. G. Barratt and M. E. Mahowald, Lecture Notes in Mathematics **657**, Springer-Verlag (1978).

[37] J. R. Harper, *Stable secondary cohomology operations*, Comment. Math. Helv. **44** (1969), 341–353.

[38] ———, *On the cohomology of stable two-stage Postnikov systems*, Trans. Amer. Math. Soc. **152** (1970), 375–388.

[39] ———, *H-spaces with torsion*, Memoirs of the Amer. Math. Soc. **223** (1979).

[40] ———, *Co-H-maps to spheres*, Israel Jr. Math. **66** (1989), 223–237.

[41] ———, *Relations in the mod 3 cohomology algebra of a space*, Bol. Soc. Mat. Mexicana **37** (1992), 203–214.

[42] J. R. Harper, H. R. Miller, *Looping Massey-Peterson towers* In: *Advances in Homotopy theory*, 69–86. S. M. Salamon, B. Steer, W. A. Sutherland (eds), London Mathematical Society Lecture Notes Series **139** Cambridge Univ. Press (1989).

[43] J. R. Harper, C. Schochet, *Coalgebra extensions in two-stage Postnikov systems*, Math. Scand. **29** (1971), 232–236.

[44] J. R. Harper, A. Zabrodsky, *Evaluating a p-th order cohomology operation*, Publicacions Matemátiques **32** (1988), 61–78.

[45] H. Hastings, *Simplicial topological resolutions*, Proc. 13th Biennial Seminar, Canad. Math. Congress **2** (1973), 66–77.

[46] J. R. Hubbuck, *Two lemmas on primary cohomology operations*, Proc. Camb. Phil. Soc. **68** (1970), 631–636.

[47] ———, *Secondary operations, K-theory and H-spaces*, In: *Workshop and Algebraic Topology Proceedings*, 79–86, Publicacions Seccio de Matemàtiques **26**, Universitat Autonoma de Barcelona (1982).

[48] R. M. Kane, *The Homology of Hopf Spaces*, North Holland Mathematical Notes **40**, North-Holland, Amsterdam (1988).

[49] ———, *Operations in connective K-theory*, Memoirs Amer. Math. Soc. **254** (1981).

[50] ———, *Implications in Morava K-theory*, Memoirs Amer. Math. Soc. **340** (1986).

[51] M. Kervaire, *A manifold which does not admit any differentiable structure*, Comment. Math. Helv. **34** (1960), 256-270.

[52] M. Kervaire, J. W. Milnor, *Groups of homotopy spheres I*, Ann. of Math. **77** (1963), 504–537.

[53] N. Kitchloo, K. Shankar, *On complexes equivalent to S^3-bundles over S^4*, International Math. Research Notices.

[54] L. Kristensen, *On the cohomology of two-stage Postnikov systems*, Acta Math. **107** (1962), 73–123.

[55] ———, *On secondary cohomology operations*, Math. Scand. **16** (1965), 97–115.

[56] ———, *On secondary cohomology operations II*, In: *Conference on Algebraic Topology*, 117–133, University of Illinois at Chicago Circle (1968).

[57] J. P. Lin, *Torsion in H-space I, II*, Ann. of Math. **103** (1976), 457–486, Ann. of Math. **107** (1978), 41–88.

[58] ———, *Two torsion and the loop space conjecture*, Ann. of Math. **115** (1982), 35–91.

[59] W.-H. Lin, *Cohomology extensions in certain 2-stage Postnikov systems*, Contemporary Mathematics **19** (1983), 177–188.

[60] A. Liulevicius, *The factorization of cyclic reduced powers by secondary cohomology operations*, Memoirs Amer. Math. Soc. **42** (1962).

[61] ———, *The cohomology of Massey-Peterson algebras*, Math. Zeit. **105** (1968), 226–256.

[62] I. Madsen, R. J. Milgram, *The Classifying Spaces for Surgery and Cobordism of Manifolds*, Ann. of Math. Studies **92**, Princeton Univ. Press, Princeton, NJ (1979).

[63] M. E. Mahowald, *On obstruction theory in orientable fiber bundles*, Trans. Amer. Math. Soc. **110** (1964), 315–349.

[64] ———, *Some Whitehead products in S^n*, Topology **4** (1965), 17–26.

[65] ———, *Some remarks on the Kervaire invariant problem from a homotopy point of view*, Proc. Sym. Pure Math. XXII, Madison, Wisc. (1970), 165–169.

[66] ———, *The index of a tangent 2-field*, Pacific Jr. Math. **58** (1975), 539–548.

[67] ———, *A new infinite family in ${}_2\pi_*^s$*, Topology **16** (1977), 249–256.

[68] M. E. Mahowald, F. P. Peterson, *Secondary operations on the Thom class*, Topology, **2** (1964), 367–377.

[69] M. E. Mahowald, M. C. Tangora, *Some differentials in the Adams spectral sequence*, Topology **6** (1967), 349–369.

[70] ———, *On secondary operations which detect homotopy classes*, Bol. Soc. Mat. Mexicana **12** (1967), 71–75.

[71] M. E. Mahowald, R. F. Williams, *The stable homotopy of $K(Z,n)$*, Bol. Soc. Mat. Mexicana **11** (1966), 22–28.

[72] W. S.Massey, *On the cohomology ring of a sphere bundle*, Jr. Math. Mech. **7** (1958), 265–289.

[73] W. S. Massey, F. P. Peterson, *The cohomology structure of certain fiber spaces I*, Topology **4** (1965), 47–65.

[74] ———, *The mod 2 cohomology structure of certain fiber spaces*, Memoirs Amer. Math. Soc. **74** (1967).

[75] C. R. F. Maunder, *Cohomology operations of the n'th kind*, Proc. London Math. Soc. **13** (1963), 125–154.

[76] J. F. McClendon, *Higher order twisted cohomology operations*, Invent. Math. **7** (1969), 183–214.

[77] J. P. Meyer, *Functional cohomology operations and relations*, Amer. Jr. Math. **87** (1965), 649–683.

[78] R. J. Milgram, *The structure over the Steenrod algebra of some 2-stage Postnikov systems*, Quart. Jr. Math. **20** (1969), 161–169.

[79] ———, *Symmetries and operations in homotopy theory*, Proc. Symp. Pure Math XXII, Amer. Math. Soc. (1971).

[80] ———, *Unstable homotopy from a stable point of view*, Lecture Notes in Mathematics **368**, Springer-Verlag (1974).

[81] J. W. Milnor, *The Steenrod algebra and its dual*, Ann. of Math. **67** (1958), 150–171.

[82] J. W. Milnor, J. C. Moore, *On the structure of Hopf algebras*, Ann. of Math. **81** (1965), 211–264.

[83] J. C. Moore, *The double suspension and p-primary components of the homotopy groups of spheres*, Bol. Soc. Mat. Mexicana, **1** (1956), 28–37.

[84] J. C. Moore, J. A. Neisendorfer, *Equivalence of Toda-Hopf invariants*, Israel Jr. Math. **66** (1989), 300–318.

[85] R. E. Mosher, M. C. Tangora, *Cohomology Operations and Applications in Homotopy Theory*, Harper and Row Publishers, New York (1968).

[86] S. Mukohda, *On a theorem of Toda in the Steenrod algebra*, Memoirs of Fac. Sci. Kyusha Univ. Ser A **14** (1960), 85–97.

[87] A. Negishi, *Exact sequences in the Steenrod algebra*, J. Math. Soc. Japan **10** (1958), 71–78.

[88] C. Pacati, P. Pavesic, R. Piccinini, *The Dold-Lashof-Fuchs construction revisited*, Rendiconti Seminario Matem. e Fisico Milano, Vol. LXV (1995) 35–52.

[89] F. P. Peterson, *Functional cohomology operations*, Trans. Amer. Math. Soc. **86** (1957), 197–211.

[90] ———, *Whitehead products and the cohomology structure of principal fiber spaces*, Amer. Jr. Math. **82** (1960), 649–652.

[91] F. P. Peterson, N. Stein, *Secondary cohomology operations: two formulas*, Amer. Jr. Math. **81** (1959), 281–305.

[92] ———, *Secondary characteristic classes*, Ann. of Math. **76** (1962), 510–523.

[93] F. P. Peterson, E. Thomas, *A note on non-stable cohomology operations*, Bol. Soc. Mat. Mexicana **3** (1958), 13–18.

[94] D. R. Ravenel, *A definition of exotic characteristic classes of spherical fibrations*, Comment. Math. Helv. **47** (1972), 421–436.

[95] ———, *The non-existence of odd primary Arf invariant elements in stable homotopy*, Math. Proc. Camb. Phil. Soc. **83** (1978), 4299-443.

[96] M. Rothenberg, N. E. Steenrod, *The cohomology of classifying spaces of H-spaces*, Bull. Amer. Math. Soc. **71** (1965), 872–875.

[97] Y. B. Rudyak, *On category weight and its applications*, Topology **38** (1999), 37–55.

[98] P. S. Selick, *Odd primary torsion in $\pi_k S^3$*, Topology **17** (1978), 407–412.

[99] ———, *A reformulation of the Arf invariant one mod p problem and applications to atomic spaces*, Pacific Jr. Math. **108** (1983), 431–450.

[100] P. S. Selick, *Introduction to Homotopy Theory*, Fields Institute Monographs **9**, Amer. Math. Soc. Providence, R.I. (1997).

[101] J.-P. Serre, *Cohomologie modulo 2 des complexes d'Eilenberg-Maclane*, Comment. Math. Helv. **27** (1953), 198–232.

[102] N. Shimada, T. Yamanoshita, *On triviality of the mod p Hopf invariant*, Japan Jr. Math. **31** (1961), 1–25.

[103] L. Smith, *Homological algebra and the Eilenberg-Moore spectral sequence*, Trans. Amer. Math. Soc. **129** (1967), 58–93.

[104] ———, *The cohomology of stable two stage Postnikov systems*, Illinois Jr. Math. **11** (1967), 310–329.

[105] ———, *Lectures on the Eilenberg-Moore spectral sequence*, Lecture Notes in Mathematics **134**, Springer-Verlag (1970).

[106] E. Spanier, *Secondary operations on mappings and cohomology*, Ann. of Math. **75** (1962), 260–282.

[107] ———, *Higher order operations*, Trans. Amer. Math. Soc. **109**, (1963), 509–539.

[108] J. D. Stasheff, *H-spaces from a Homotopy Point of View*, Lecture Notes in Mathematics **161**, Springer-Verlag (1970).

[109] N. E. Steenrod, *Cohomology Operations*, Lectures by N. E. Steenrod. Written and revised by D. B. A. Epstein. Ann. of Math. Studies **50**, Princeton Univ. Press, Princeton, N.J. (1962).

[110] ———, *A convenient category of topological spaces*, Mich. Math. Jr. **14** (1967), 133–152.

[111] H. Toda, *Composition methods in homotopy groups of spheres*, Ann. of Math. Studies **49**, Princeton Univ. Press, Princeton, N.J. (1962).

[112] ———, *Extended p-th powers of complexes and applications to homotopy theory*, Proc. Japan Acad. **44** (1968), 198–203.

[113] E. Thomas, *Seminar on fiber spaces*, Lecture Notes in Mathematics **13**, Springer-Verlag (1966).

[114] ———, *Postnikov invariants and higher order cohomology operations*, Ann. of Math. **85** (1967), 184–217.

[115] ———, *Whitney-Cartan product formula*, Math. Z. **118** (1970), 115–138.

[116] ———, *The Thom-Massey approach to embeddings*, In: *The Steenrod Algebra and its Applications*, 283–317, F. P. Peterson, ed. Lecture Notes in Mathematics **168**, Springer-Verlag (1970).

[117] G. W. Whitehead, *Elements of Homotopy Theory*, Springer-Verlag (1978).

[118] A. Zabrodsky, *Implications in the cohomology of H-spaces*, Illinois Jr. Math. **14** (1970), 363–375.

[119] ———, *Secondary cohomology operations in the module of indecomposables*, Summer Inst. on Algebraic Topology, Aarhus (1970).

[120] ———, *Secondary operations in the cohomology of H-spaces*, Illinois Jr. Math. **15** (1971), 648–655.

[121] ———, *The classification of H-spaces with three cells I, II*, Math. Scand. **30** (1972), 193–210, 211–222.

[122] ———, *Some relations in the mod 3 cohomology of H-spaces*, Israel Jr. Math. **33** (1979), 59–72.

[123] ———, *On the realization of invariant subgroups of $\pi_*(X)$*, Trans. Amer. Math. Soc. **285** (1984), 467–496.

[124] ———, *Cohomology Operations and H-spaces*, Publicacions, Seccio de Matematiques, Univ. Autonoma de Barcelona, **26** (1982), 215–243.

[125] ———, *Hopf Spaces*, North-Holland Math. Studies **22**, North-Holland, Amsterdam (1976).

Index